Wireless Mesh Networks

Wireless
Mesh
Networks

Gilbert Held

Auerbach Publications
Taylor & Francis Group

Boca Raton   London   New York   Singapore

Published in 2005 by
Auerbach Publications
Taylor & Francis Group
6000 Broken Sound Parkway NW, Suite 300
Boca Raton, FL 33487-2742

Library of Congress Cataloging-in-Publication Data

Held, Gilbert, 1943-
   Wireless mesh networks / Gilbert Held.
      p. cm.
   ISBN 0-8493-2960-4 (alk. paper)
    1. Wireless communication systems. 2. Routers (Computer networks). I. Title.

TK5103.2.H453 2005
621.382'15--dc22   2005041065

Taylor & Francis Group
is the Academic Division of T&F Informa plc.

Visit the Taylor & Francis Web site at
http://www.taylorandfrancis.com

and the Auerbach Publications Web site at
http://www.auerbach-publications.com

Dedication

Over the past decade I found that teaching graduate school is a wonderful way to spend some evenings during the week. Although the job of a professor is to transfer knowledge, many times it's a two-way street, with thought-provoking student questions making me realize that the wonderful field of communications technology is far from reaching its promoted ability to link mankind. Thus, I would be remiss if I did not take the opportunity to thank the students at Georgia College and State University for their inquisitive minds that make teaching most interesting.

Table of Contents

Preface

Approximately every five or ten years we are blessed with the implementation of a technology that can have a major bearing upon how we work, facilitates our productivity, and even enhances our recreational capability. In the past we witnessed the PC revolution, the advent of the PDA, and the growth in the use of wireless LANs. Today a new technology referred to as wireless mesh networking has the potential to considerably influence how we communicate. Although derived from military research into mobile networks, the emergence of wireless mesh networking has its greatest potential in the commercial marketplace and is the primary focus of this book.

Within a few years, wireless mesh networking may revolutionize the manner by which you can access the Internet as well as communicate with co-workers and friends. Although the software behind wireless mesh networking is still evolving, the concept behind the technology that eliminates the need for a centralized control mechanism is well thought out and will remain in place. In this book we examine the concept behind wireless mesh networking, its advantages over existing technologies and its existing and potential applications, and explore how some of the networking protocols associated with this technology operate by obtaining an appreciation for the technology in the office, government agencies, and in a campus environment as well as in the home.

Because it's the job of an author to fully inform readers of all sides of an issue, we note that there are some significant problems associated with wireless mesh networking. Some problems are associated with the scale of the network, with an increased area of coverage requiring more stations than a smaller network. Other problems, such as security, represent issues that must be considered regardless of the size of the mesh network. Still other problems, such as radio frequency interference, can represent both controllable and uncontrollable issues because it may be difficult or impossible to control the use of machinery and fluorescent lighting by other organizations, yet alone the periodic sunspots that radiate hundreds of millions of miles onto our small planet. Thus, at applicable locations in this book we note the problems associated with wireless mesh networking as well as actual and potential solutions to such problems.

As a professional author I truly welcome reader comments. Let me know if you feel I should expand upon a topic, if I provided too much information, and what topics you might like to read about in a future edition of this book. Of course, any other comments or suggestions are also welcomed. You can contact me through my publisher whose address is on the cover of this book or you can send me an e-mail directly to gil_held@yahoo.com.

Gilbert Held
Macon, Georgia

Acknowledgments

The preparation of a book in many respects is similar to a sport in that it is a team effort. This book is certainly no exception, as it required the efforts of many people to publish the book you are now reading.

Once again this author is indebted to Richard O'Hanley at Auerbach Publications for green-lighting another idea and providing backing for this project. I would be remiss if I did not also thank the production staff at Auerbach Publications for turning this author's manuscript into the book you are reading. Concerning the manuscript, it is with a great sense of pride that I wish to thank my wife, Beverly Jane Held, for her efforts in turning my handwritten notes into a professionally typed manuscript. Beverly typed this author's first book on a 128-kb Macintosh many years ago. Although technology has certainly changed over the years, Beverly's typing skills continue to maintain a level of accuracy that is truly appreciated.

Chapter 1

Introduction to Wireless Mesh Networking

The purpose of an introductory chapter is to provide readers with basic information concerning the subject of a book. Because this book is about wireless mesh networking, as you might surmise the goal of this introductory chapter is to become familiar with this topic, terms associated with this topic, and even a few associated abbreviations.

In this chapter we begin with an explanation of mesh networking and wireless mesh networks. This explanation includes a brief examination of different types of networking, with this author discussing networking structures commonly referred to as networking topology and the manner by which such structures evolved. As we review different types of networking topologies, we note some of the advantages and disadvantages associated with each structure, which will provide a foundation for examining the advantages and disadvantages associated with wireless mesh networking. Because modern mesh networks are built upon over the-air transmission and primarily use existing wireless LAN networking components, it will come as no surprise that we also focus our attention upon this area in this chapter.

Once we complete our initial examination of networking topology, we turn our attention to the different types of mesh networks, their advantages, and disadvantages.

In concluding our introduction to the topic of mesh networks and wireless mesh networking, we examine some of the existing and evolving applications that have the potential to make wireless mesh networking into an ubiquitous technology. That said, let's grab a Coke, Diet Pepsi, or another drink and our favorite munchies and explore the wonderful world of wireless mesh networking.

1.1 Mesh Networking Defined

To understand mesh networking, we first need to obtain an appreciation for what a mesh topology represents. If we have n nodes in a network, where the term "node" refers to a communications device that can transport data from one of its interfaces to another, then the ability of each node to communicate with every other node in the network represents a mesh network topology. We can view the structure of a mesh network by simplifying the number of nodes in the network from a value of n, which is what mathematicians like to work with, to an easy-to-visualize number, such as three, four, or five.

Nodes and Links

Figure 1.1 illustrates three-, four-, and five-node mesh network structures, in which each node has a communications connection to all other nodes in the network. The connection between each node is referred to as a link.

If we examine the number of links associated with each network shown in Figure 1.1, it's obvious that the number of links increases as the number of nodes increases. Although only three links are required to interconnect three nodes, six are required to interconnect four nodes, and ten are required to interconnect five nodes. If you take the time to draw six nodes and interconnect each, you would then note the need for fourteen links. What this means is that a classical mesh network structure in which each node is interconnected to every other node in the network becomes impractical as the number of nodes in the network increases. After all, when networks were first constructed, the links interconnecting nodes were dedicated or leased telephone lines. This meant that a separate physical interface was required by a node to connect to each link. That interface primarily performed parallel to serial and serial to parallel conversion, because data flows bit by bit on a serial link. Because each interface requires buffer memory and a node is a computing device with a finite amount of processing power, adding interfaces increases the amount of processing the node needs to perform until one interface too many is added that saturates the processing capability of the node. Thus, from a classical perspective, a mesh network in which every node can directly communicate with every other node has physical constraints that limit the number of nodes that can be interconnected.

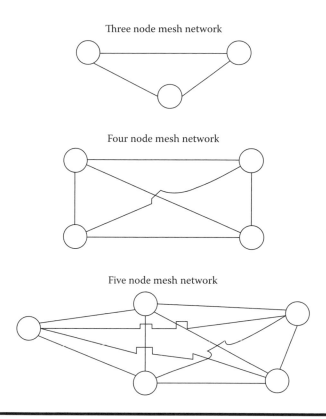

Three node mesh network

Four node mesh network

Five node mesh network

Figure 1.1   In a true mesh network structure, each node has a connection to every other node in the network.

Control Issues

Recognizing the previously mentioned constraints associated with network nodes resulted in the development of more cost-effective partial mesh network structures. A good example of a partial mesh network would be the public packet networks constructed by Tymnet and Sprint during the 1970s and 1980s. Such networks consisted of hundreds to thousands of nodes, however, instead of each node being directly interconnected to every other node, they simply had two or more links to other nodes to provide an alternate routing and traffic balancing capability. Because nodes are not directly connected to one another, traffic would typically flow through one or more intermediary nodes to its destination, requiring the development of routing protocols that are based upon the transfer of control messages between nodes. Similarly, alternate routing and traffic balancing operations also required coordination, with control messages transmitted between nodes and a centralized network operations center required for controlling the flow of messages between nodes, which in turn controlled the use of alternate links for backup and traffic balancing operations. This concept of the use of a partial mesh holds true for the "mother of all interconnected networks" better known as the Internet.

Modern Mesh Networking

In a wireless environment, a single radio frequency (RF) transmitter/receiver in one node has the ability to communicate with a virtually unlimited number of other nodes. Thus, the physical constraints associated with wired connectivity becomes less of an issue in a wireless environment. This means it's both a practical and a relatively simple process for one node to communicate with many other nodes because a single interface in the form of an RF transmitter/receiver can be substituted for the multiple interfaces required in a wired environment. Obviously, other nodes must be within transmission rage for communications to occur.

Wireless Networking Structures

There are two basic types of wireless LAN networking structures, referred to as peer-to-peer and infrastructure. In a peer-to-peer networking structure, each node can directly communicate with every other node, assuming they are in transmission range of one another. In an infrastructure wireless LAN networking environment, all traffic flows through an access point (AP). The access point represents a two-port bridge, with one port connected to a wired network and the second port representing the RF transmitter/receiver. Thus, in an infrastructure wireless network two nodes communicating with each other do so by first transmitting to the access point which then regenerates the data. Because the access point in effect functions as a relay station, when transmission occurs between two wireless nodes the transmission distance between nodes can double in comparison to a peer-to-peer networking environment. However, the access point represents a central control mechanism and, if it fails or if a node is out of range of the access point, communications suffer.

Overcoming Transmission Distance Limitations

A solution to the need for centralized control and transmission range limitations of wireless nodes occurs by enabling each network node to function as a relay. Figure 1.2 provides a comparison of a wireless peer-to-peer network, a wireless LAN infrastructure network, and wireless mesh networking. Note that in a wireless mesh networking environment each node functions as a router and repeater, forwarding data to the next node

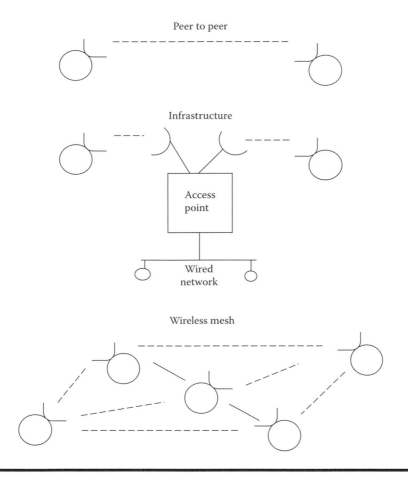

Figure 1.2 Comparing wireless LAN topologies.

toward its ultimate destination. In comparison, in a peer-to-peer environment transmission is limited to two nodes communicating with each other
whereas in an infrastructure networking environment all transmissions
occur through a centralized access point. However, because nodes can be
modified to relay information we can group a sequence of peer-to-peer
transmissions to obtain a mesh structure operating environment. Based
upon the preceding, we can define a wireless mesh network as follows.

> *A wireless mesh network represents a series of peer-to-peer trans-
> missions where each node functions as a router and repeater.*

Note from the above definition that there is no requirement for any
centralized control and in fact nodes communicate with each other on a
peer-to-peer basis.

We can use an analogy to obtain additional information concerning the operation of a mesh network. Assume a room has a group of people, each having a similar communications device, such as a cell phone or notebook computer. If we are located in the room and need to send a message to a person located outside the room we could either move through the room or convey our message to another person, with the expectation that that person would relay our message until a person near the door could open it and deliver the message to the appropriate person located outside the room. If we think of each person in the room as a "client" or "node" then one person can relay our message to another as a peer-to-peer transmission. The person who opens the door to communicate our message to the person behind the door can be thought of as a network gateway. Thus, in a mesh networking environment, messages are passed in the form of electronic signals from client to client or node to node until they reach their destination on the network or a gateway for transmission off the mesh.

Now that we have a general appreciation for the term "wireless mesh networking," let's backtrack a bit and discuss the general aspects of how networks evolved. Doing so provides us with a firmer understanding of wireless mesh networking including its relationship to wireless LANs and the advantages and disadvantages associated with the technology.

1.2 Network Evolution

Earlier in this chapter when we examined mesh networking we noted the term "link" was used to refer to the connection between two nodes. Both of those terms, "link" and "node," are considered by some people as antiquated in today's modern wireless environment. However, they represent a good starting point for examining the evolution of networking including different types of networks and mechanisms to move or transport data from one network to another.

Network Topologies

There are two generic types of network topologies or structures that evolved over the past half century. Those two generic topologies are referred to as point-to-point and multipoint.

Point-to-Point

The first type of network structure consisted of a link interconnecting two nodes. This network structure is simply referred to as a point-to-point link

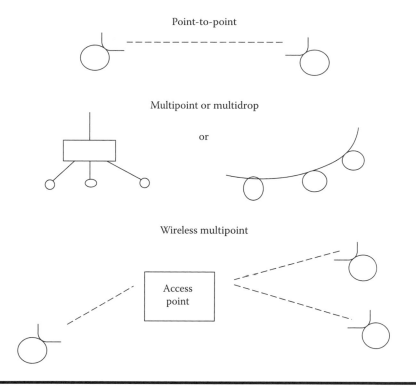

Figure 1.3   Generic network topologies.

because it directly interconnects two locations. The top portion of Figure 1.3 illustrates a point-to-point network structure.

Multipoint

To conserve the cost of leased lines, vendors developed poll and select software that enabled multiple terminal devices to be connected to a common communications line. Referred to as multidrop or multipoint networking, terminals were either individually located in different geographical areas or were clustered together and connected to a common communications line by a control unit. For either situation poll and select software permits multiple terminal devices to share a common communications line. The middle portion of Figure 1.3 illustrates the two types of multipoint communications used in a wired network environment.

In a wireless environment the transmission of data from an access point can also be considered as multipoint communications. Although poll and select software is not necessary, the access point must address data to specific devices with which it wishes to communicate. Similarly each device that needs to communicate with an access point must use the destination address of the AP (Access Point) to ensure data is delivered correctly. The lower portion of Figure 1.3 illustrates an example of wireless multipoint communications.

Through the use of a central facility referred to as a hub, you can create a star topology by routing point-to-point lines into the hub. Similarly, creating drops off a common point-to-point line creates what is referred to as a bus network structure. The key to the ability to create various network structures resides in different types of network addressing. Thus, let's turn our attention to this topic.

Types of Networking Addressing

There are three basic types of networking addresses, technically referred to as unicast, multicast, and broadcast.

Unicast Addressing

In a unicast address networking environment a packet has a single destination address. One of the earliest types of network architectures involved the connection of a front-end processor to control units, with each control unit interfaced to a group of terminals. In this networking environment each control unit represented a node and each connection between the front-end processor and a node represented a link. Data flowing from the front-end processor to an individual terminal connected to a control unit needed a specific terminal address to reach its intended destination. That address, which represented the specific terminal, became known as a unicast address.

Broadcast Addressing

If it became necessary to transmit a message to every terminal connected to a control unit, the flow of a sequence of messages in which only the destination address changed would be far from efficient. Recognizing this fact resulted in the development of a broadcast address. In our front-end processor to control unit configuration example, the control unit would examine each transmission from the front-end processor. If the destination address was for a specific terminal connected to the control unit the data would then be directed to that terminal. In comparison, if the destination address was a control unit broadcast address, the control unit would then replicate the message to all attached terminals.

In a modern client/server environment routers replaced front-end processors and control units. Although each router functions as a network node the term "node" is rarely used today for this type of device. In addition, because most routers are connected to LANs, which in turn provide connectivity for tens to hundreds or thousands of personal computers, the need for a mechanism to transmit a common message to all computers on the network in an efficient manner continues. Because routers operate at the network layer whereas frames flowing on a LAN operate at the data link layer, this resulted in two types of broadcast addresses: a network (layer 3) broadcast address and a media access control (layer 2) broadcast address.

Types of Broadcast Addresses

In an Internet Protocol (IP) networking environment a network broadcast address is formed by setting the value of the host address portion of the Class A, B, or C address to all 1s. For example, if the network address is 198.78.46.0, which is a class C network address, the broadcast address for that network then becomes 198.78.46.255. Here the value of 255 represents the decimal value associated with setting each bit in the last byte of the address to a value of 1. Table 1.1 lists the broadcast address values for Class A, B, and C IP networks, where the entries x. x. x and x. x. x represent valid dotted decimal values for each applicable IP class address.

Although a broadcast IP network address is used to deliver one copy of a packet to all stations on a network, the actual delivery of the packet requires an address translation. This is because packets are delivered on a LAN via a data link layer 2 protocol, such as Ethernet or Token Ring. Thus, the layer 3 IP broadcast address must be converted into a layer 2 broadcast address. This occurs by the router encapsulating the IP packet into a LAN frame and filling in the 48-bit MAC destination address of the frame with all 1s.

Although a network broadcast address needs to be converted into a layer 2 broadcast address for delivery, the reverse is not true. That is, stations connected to a LAN can transmit broadcast frames to each other without the frames having to be converted into IP packets. In fact, at layer 2 broadcasts are restricted to individual LANs. Now that we have an appreciation for unicast and broadcast transmission let's turn our attention to a third type of transmission which is referred to as multicast and which is based upon multicast addressing.

Table 1.1   IP Broadcast Addresses

Class A	x.255.255.255
Class B	x.x.255.255
Class C	x.x.x.255

Multicast Addressing

A multicast address represents a special type of group address. To understand the rationale for multicast addressing assume there are ten client stations on a LAN whose operators wish to receive a video of the latest Mars landing. Without multicast addressing each operator would download an individual copy of the video in the form of a unicast transmission. Thus, ten individual video broadcasts would flow over the Internet and onto the LAN where the ten client stations reside. In addition to consuming valuable Internet resources in the form of having routers allocate processing power to groups of packets that only vary by their destination address, the multiple packets would consume both Internet and LAN bandwidth. Recognizing the potential waste of resources when multiple clients require access to the same data flow was the key motivation for multicast addressing.

Under multicasting, a term used to denote the multicast process, clients subscribe to an event, such as the previously mentioned Mars landing video. In effect, the subscription results in the client station's software being programmed to receive all frames that have a destination address associated with a specific multicast. Then, one sequence of packets flows through the Internet and onto the LAN, with each client that is associated with the multicast pulling copies of the LAN frames into memory for processing. As you might expect, other clients on the LAN that are not members of the multicast simply ignore LAN frames that are part of the multicast. Thus, multicast transmission represents a mechanism that conserves router processing power as well as Internet and LAN bandwidth. Now that we have an appreciation for the three types of network transmission and network addressing, let's focus our attention upon the two major methods used to move data from one network to another: bridging and routing.

Bridging and Routing

Bridging represents a layer 2 method for controlling the flow of data. Operating at the Media Access Control (MAC) layer, bridging is a relatively simple process for moving data that depends upon what are referred to as the three Fs: flooding, forwarding, and filtering.

A bridge operates by constructing what is referred to as a port-address table. As a frame enters a port on the bridge it examines its destination address and compares that address to entries in its port-address table. If the bridge cannot locate a matching destination address it floods the frame onto all ports other than the port where it entered the bridge. In addition, the bridge notes the source address of the frame and the port where it entered the bridge, using this information to update its port-address table.

If the bridge can match the destination address in the frame with an address in its port-address table it will normally forward the frame onto the port associated with the address, a technique referred to as forwarding. The only exception to forwarding is when the port in the port-address table matches the port on which the frame was received. Because it makes no sense to forward a frame back to where it came from, the bridge filters the frame, in effect placing it into the big bit bucket in the sky. Thus, a bridge operates based upon the 3 Fs, flooding, forwarding, and filtering, using layer 2 MAC addresses as decision criteria.

Routing represents a layer 3 method for controlling the flow of data. Because layer 3 represents the network layer, routing involves the flow of data between networks. In comparison, bridging occurs based upon layer 2 addressing, which results in the control of data flow between subnets linked together by a bridge.

There are two key differences between bridging and routing. First, bridges are self-learning devices whereas routers require manual configuration. This means that a bridge is essentially a plug-and-play device that can be removed from its packing carton and installed without requiring manual configuration. In comparison, as a minimum, routers need to have their layer 3 addresses defined for each interface as well as other configuration data either entered into the device or selected from a predefined configuration list of options. Because routers are capable of traffic balancing and alternate routing they require a mechanism to alter the flow of data through a network, which results in the second major difference between bridges and routers. Routers employ a routing protocol, which enables data to flow from source to destination between intermediate devices that can provide alternate routing in the event traffic clogs a more favorable path or in the event an existing path becomes inoperative. In comparison, bridges simply examine layer 2 MAC addresses and either forward, flood, or filter packets without any ability to consider alternate paths for the flow of data. Now that we have a general appreciation for the differences between bridging versus routing, let's focus our attention upon the topological characteristics of wireless LANs.

Wireless LAN Topology

Every participant on a wireless LAN is commonly referred to as a station. Because an access point and a client connected to an access point can both be referred to as stations, the terms "client" and "access point" are better suited to distinguish one type of device from another and are used in this book. That is, client refers to laptops, desktops, and PDAs equipped with wireless LAN adapter cards, and access point is used to refer to a bridge that provides access between wireless and wired network devices.

Using the preceding information, let's turn our attention to the manner by which groups of devices communicate with one another and the wireless terminology used to refer to certain groupings of devices.

Service Sets

The grouping of two or more wireless LAN devices results in the formation of a service set. The actual type of service set that is formed depends upon the type of wireless LAN devices and the manner by which they communicate. When two or more clients communicate directly with each other (known as peer-to-peer or ad hoc communications), they form what is referred to as an independent basic service set. In comparison, when one or more clients communicate via the use of an access point, the AP and clients form an infrastructure service set. Because both an independent service set and an infrastructure service set have the same initials (ISS), we do not use these initial abbreviations due to possible confusion.

Basic Service Set

The term "Basic Service Set" (BSS) can be used to refer to any group of two or more wireless devices that communicate with each other, such as two or more clients that form an independent basic service set or one or more clients and an access point that form an independent basic service set. For either situation the wireless devices have a limited range. Thus, to enable wireless LAN devices to communicate with one another at extended distances required a mechanism to convey information between two or more BSSs. That mechanism is referred to as a Distribution System (DS) which is used to interconnect two BSSs that can be in the same building or located on different continents.

Distribution System

The actual medium used for the DS is not defined. Thus, the DS could be a wired Ethernet LAN, a point-to-point leased line, or even a wireless repeater. Because the BSSs are in effect extended by the DS, the connection of two BSSs by a DS is referred to as an Extended Service Set (ESS). Figure 1.4 illustrates the relationship between two BSSs and a DS that forms an ESS.

Mesh Network Evolution

Although wireless mesh networks can be based upon a variety of technologies, their practical commercial evolution is primarily occurring

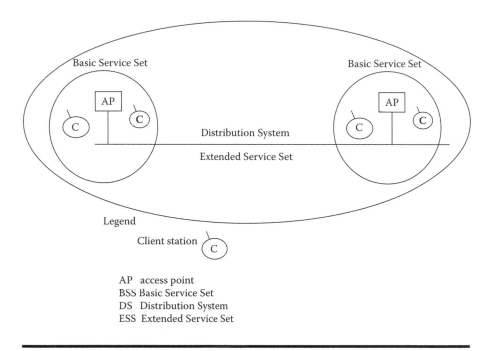

Figure 1.4   Relationship of wireless LAN service sets.

through the use of wireless LAN communications. One of the first types of networks that evolved into a wireless mesh network structure was the wireless LAN ad hoc network. As previously discussed in our review of wireless LAN topology, the wireless peer-to-peer network structure is also referred to as an ad hoc network. In addition, an ad hoc network is commonly referred to as an infrastructureless network inasmuch as client stations communicate directly with other clients instead of having to go through a centralized access point. Because clients can enter and exit the network at will, the term ad hoc is also used to describe this type of networking environment.

Because an ad hoc network lacks a centralized authority, each node needs to be capable of relaying information for data to move between nodes. The top portion of Figure 1.5 illustrates a four-node ad hoc network, indicating how packets can be routed from node A to D. In this example, data is not transmitted directly from source to destination. Instead, data is forwarded through intermediate nodes that in effect act as routers. If the top portion of Figure 1.5 represented a wired network, every node except the endpoints would contain two ports. The routing tables would be similar to those used by bridges, with the address of devices associated with the ports on each node. The lower portion of Figure 1.5 illustrates

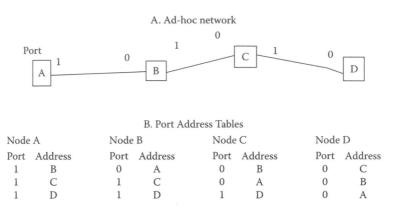

Figure 1.5 **An ad hoc network where nodes provide routing.**

the port-address routing tables that would be established through the bridge learning process previously described in this chapter.

The network shown at the top of Figure 1.5 is formed by the routing information developed in each node. Thus, an entry in node A's routing table informs it that B is the next hop for all packets destined to nodes B, C, and D. In a wireless environment each node has a single port in the form of an antenna. If a directional antenna is used, it becomes possible for the beam width and focus of the antenna to be adjusted based upon routing table entries.

Routing Algorithms

There are two routing algorithms that are commonly used with ad hoc networks that form the basis for the development of a mesh network. Those routing algorithms are dynamic source routing and on-demand distant vector. Thus, let's briefly discuss how each algorithm operates, with a more detailed examination to occur late in this book.

Dynamic Source Routing

Dynamic source routing represents one of several protocols being investigated by the Internet Engineering Task Force (IETF) for use by ad hoc wireless networks. This routing protocol is based upon the concept of source routing, but it was modified to enable each node to be mobile. As a refresher, source routing represents an Internet Protocol (IP) option. When enabled, this option allows the originator of a packet to specify the path it will take to its destination as well as the path responses take when the destination responds to the originator. Source routing is defined in the RFC791 standard and is primarily used for diagnostic testing of a specified route or when the default route for a connection is not optimal, perhaps due to an expected rise in traffic resulting from the availability of a new service.

Dynamic source routing is similar to IP source routing. Under dynamic source routing, a route request is used to determine the path from the source to the destination. The destination issues a route reply, which provides the reverse path. Although the route between source and destination does not need to be a reverse image of the path between destination and source, some protocols require bidirectional connections. One such protocol is the IEEE 802.11 standard, which enables a destination station on a wireless LAN using dynamic source routing to simply reverse the route to itself to determine the route to the source. In a dynamic source routing environment, each node examines every packet it receives, an operating method referred to as promiscuous mode. As the node examines the addresses in each packet, it learns where other devices are located relative to the node examining packets. Due to this, nodes do not need to transmit periodic routing advertisements, such as Routing Information Protocol (RIP) transmissions that are used to inform other nodes of the state of the network.

On-Demand Distant Vector

A second ad hoc routing algorithm is the On demand Distance Vector (ODV) algorithm. Under this routing algorithm, each router advertises its view of the network to its neighbors in the form of subnets directly connected to the device. Each neighbor uses this information to compute its distance to all other subnets. Unlike dynamic source routing that avoids routing advertisements, the on-demand distance vector algorithm uses periodic "HELLO" messages to track the state of the link between two nodes.

Ad Hoc Mesh Networks

Now that we have an appreciation for the two routing protocols primarily used in ad hoc networks, let's go a step further and examine how such networks can form the basis or foundation for a mesh network. As we previously noted, an ad hoc network consists of a series of two or more nodes communicating as peers. As we add additional nodes to the network, we obtain a mesh structure. The mesh structure represents a multidrop system in which nodes assist other nodes in transmitting packets through the network. Each node functions as a router, relaying packets for its neighbors. Through the relay process, a packet will be forwarded through intermediate nodes to its destination. In addition, because the routing protocol will adjust to its environment, nodes can enter and leave the network. Because modern-day communications are primarily client/ server, with the server representing a Web or mail server, the mesh network will more than likely be connected to the Internet. This is accomplished by connecting one of the nodes in the mesh to a router which in turn is connected to the Internet. Although additional nodes can be added to the network as long as they support the same routing algorithm used by other nodes, at a certain point in time the searching of routing tables will hinder network performance. Thus, there is obviously an upper limit on the number of nodes that can be grouped to form a mesh network. However, the upper limit will vary based upon the use of the network, with some applications that require less transmission delay having a lower limit on nodes than do other applications. Now that we have a general appreciation for routing protocols and the formation of a mesh network via a series of ad hoc nodes, let's turn our attention to the advantages afforded through the use of a mesh network.

Advantages of Use

There are many reasons to consider the use of a wireless mesh network. Some reasons, as we note shortly, may be more appropriate to one type of operating environment than another. The top portion of Table 1.2 lists six common advantages associated with the use of a wireless mesh network.

Reliability

In a wireless mesh network each node functions as a relay to move packets toward their ultimate destination. Because nodes can enter and leave the mesh, each node must be capable of dynamically changing its

Table 1.2   Advantages and Disadvantages Associated with the Use of Mesh Networks

Advantages	Disadvantages
Reliability	Lack of standards
Self-configuration	Security
Self-healing	Overhead
Scalability	
Economics	

forwarding pattern based upon its neighborhood. Thus, the mesh topology enhances reliability because the failure of one link due to RF interference, the movement of a vehicle between source and destination or another phenomenon, will result in packets being forwarded via an alternative link toward their destination.

Self-Configuration

Because nodes in a mesh network learn their neighbors and paths to other nodes, there is no need to configure each node. Thus, the self-configuration capability of nodes can considerably reduce the need for network administration.

Self-Healing

Because nodes in a wireless mesh network dynamically learn their neighbors as well as links to other nodes, there is automatic compensation for the failure or removal of a node. Thus, in the event of a transmission impairment that adversely affects the use of a link or the failure of a node, other nodes establish alternate paths. The establishment of alternative paths results in a self-healing capability.

Scalability

As we previously noted, nodes can enter and exit a mesh network as long as they operate software compatible with other nodes in the network. This means that you can extend the area of coverage of a wireless mesh network by simply placing new nodes at appropriate locations where they can communicate with existing network nodes. Thus, a wireless mesh network is scalable. However, the number of nodes you may require and the upper limit concerning the total number of nodes you can have in a network will vary based upon your organization's physical and technical operating environment. Concerning the physical environment, the number of obstacles as well as the level of RF interference will govern the number of nodes required within a given area. In comparison, the routing protocol and processor capability of each node will govern the number of nodes that can be added to a network prior to transmission degrading to an unacceptable level.

Economics

If we consider some of the previously mentioned benefits associated with wireless mesh networks, one implied but not discussed additional benefit is economics. Because a wireless mesh network does not require centralized administration nor do nodes require manual configuration, such networks are less expensive to set up and operate than conventional networks. Similarly, the ability of wireless mesh networks to automatically resolve link and node outage problems via their self-healing capability eliminates the necessity for manual intervention when things go wrong. Because employee hourly rates can easily exceed the cost of wireless LAN adapters and even some access points, the elimination or even a reduction of the need for manual intervention can provide considerable economic benefits for most organizations.

1.3 Disadvantages of Use

To ensure readers have a fair and balanced view of wireless mesh networks, this author would be remiss if he did not mention some of the disadvantages associated with the technology. Those disadvantages which are listed in the lower portion of Table 1.2 include a lack of standards, security, and the overhead associated with relaying packets through nodes to their ultimate destination.

Lack of Standards

At the time this book was written, several organizations were in the process of developing wireless mesh networking standards. Because it will be several years until such standards are promulgated and supported by vendor equipment, organizations that currently create mesh networks can be viewed as pioneers. Although some people may view pioneers as those susceptible to catching arrows in their backs, creating a wireless mesh network using proprietary software is not necessarily bad. The key factor to note is the lack of interoperability between different vendors because there are no existing standards with which vendors can tailor their products to comply.

Security

Because nodes within a wireless mesh network function as routers relaying packets to other nodes, security is an important issue. As the number of nodes in a wireless mesh network increases, you in effect have more locations where insidious persons can view your data. In addition, if software permits nodes to be added without centralized control, a mechanism is required to ensure the node is legitimate and not a PC operated by a hacker. This means that a method of authentication of nodes is required in addition to securing the flow of data through nodes, two security areas that need to be addressed by both standards and proprietary products.

Overhead

The old adage, "There is no free lunch," is also applicable to wireless mesh networks. Because nodes must learn their neighbors as well as paths to other nodes, they must create and maintain routing tables. As network traffic and the number of nodes in the network increases, so will the amount of processing devoted to routing packets. Thus, the efficiency of routing software as well as the number of network nodes and level of network traffic results in processor overhead that can adversely affect the performance of the node to perform other tasks. Now that we have an appreciation for the advantages and disadvantages associated with wireless mesh networks, we conclude this introductory chapter with a brief examination of potential applications suitable for this technology.

1.4 Applications

Similar to other communications technologies, wireless mesh networks can support a variety of applications that is only limited by one's imagination. Currently, the primary use of the technology is to support extended access to the Internet within a geographical area.

Over the past year several vendors introduced proprietary products that through software convert IEEE 802.11 client stations into mesh network nodes capable of forwarding packets from other nodes toward a gateway. The gateway, which is connected to the Internet, provides all nodes with Internet access.

If you envision a group of suburban homes or a group of apartments within a building, you can immediately visualize the economic benefits afforded by a wireless mesh network. Instead of each homeowner or apartment dweller having a separate cable or DSL modem connection to the Internet, the use of mesh networking technology can considerably reduce Internet connections. Depending upon the number of homeowners or apartment dwellers, perhaps one or a few Internet connections will be able to replace several dozen to a hundred or more separate connections.

In a business environment wireless mesh networking can be used to extend wireless LANs between floors within a building and even as a mechanism to interconnect two or more buildings. Although it's easy to visualize the economic benefits afforded by reducing the number of Internet connections from one per client to one per group of clients, redundancy, reliability, and scalability are also enhanced. In addition, because a mesh network could provide multiple paths to the Internet that are dynamically adjusted when transmission impairments occur, the redundancy of such networks will provide enhanced reliability. This means that a wireless mesh network can be suitable for factory floors and other "harsh" RF environments where current networking technology encounters limitations due to the environment.

Chapter 2

Radio Frequency Utilization

In the first chapter in this book we became acquainted with various networking terms including obtaining a definition of a mesh network and learning about the service sets and operational modes of wireless LANs. In this chapter we turn our attention to obtaining an understanding of how wireless mesh networks utilize the frequency spectrum. In doing so we become familiar with such terms as "frequency" and "bandwidth," the location in the frequency spectrum where different frequency bands used by wireless LAN networks reside, and power measurements, because the latter are necessary to obtain an understanding of the relationship between antenna sensitivity and transmission range. Using power measurement information then allows us to discuss the use of antennae and how antenna sensitivity and barriers in the form of buildings, vehicles, and even people affect transmission range. Thus, this chapter can be viewed as building upon our prior knowledge of mesh networking presented in the previous chapter by focusing our attention upon some specific technical information that governs the operation of wireless transmission.

2.1 Frequency, Period, and Bandwidth

Three key terms that govern the operation of wireless devices are frequency, period, and bandwidth.

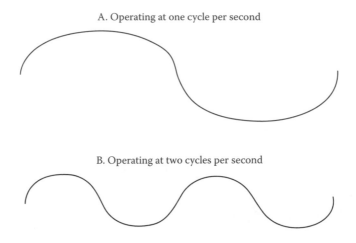

Figure 2.1   The sine wave.

Frequency

Frequency represents the oscillation or movement of a signal per unit of time. In the field of data communications the sine wave is commonly used as a signal onto which information is impressed via modulation, with the term modulation representing the altering of a signal. Thus, to obtain an appreciation of frequency, let's focus our attention upon the sine wave.

The top portion of Figure 2.1 indicates a sine wave operating at one (1) cycle per second. The complete signal cycle is shown to require one second; hence, we say the frequency of the signal is one cycle per second or 1 cps. Another term used to express frequency is Hertz (Hz) in honor of the German physicist, where 1 Hz is equivalent to 1 cps.

The lower portion of Figure 2.1 illustrates a sine wave operating at 2 cps or 2 Hz. Note that at 2 cps the sine wave completes two cycles in one second. Similarly, if the sine wave made three complete cycles in one second its frequency would be 3 cps or 3 Hz, and the occurrence of four cycles in one second would result in a frequency of 4 cps or 4 Hz. Thus, as the number of cycles per period of time increases, the frequency of the oscillations increases.

Period

As we just noted, the frequency of a wave refers to how often it completes a cycle with respect to a uniform period of time. That period of time is normally expressed as a second because the terms cps and Hz refer to oscillations within that amount of time.

The relationship between frequency and period is shown below.

$$\text{Frequency} = 1/\text{period}$$
$$\text{Period} = 1/\text{frequency}$$

As noted, frequency and period are inversely proportional to each other. We can consider frequency as representing how often something happens, whereas period represents the time in which the event occurred. Because the symbol f is used for frequency and the symbol T is used for period, the relationship between frequency and period can also be expressed as follows.

$$F = 1/T \quad \text{and} \quad T = 1/F$$

Now that we understand the relationship between frequency and period, let's turn our attention to a third key term, bandwidth.

Bandwidth

Bandwidth represents a range of frequencies. Normally, when we discuss bandwidth requirements of a communications device we refer to a contiguous range of frequencies required by the device. Thus, we can express bandwidth as follows.

$$B = f_H - f_L$$

where
   f_H = highest frequency
   f_L = lowest frequency

Frequency Bands

In just about every country on our globe, the use of the frequency spectrum is regulated. To prevent regulations from stifling the development of electronic equipment most countries belong to the International Telecommunications Union (ITU) and comply with its recommendations concerning the utilization of frequency. Doing so enables such varied communications-dependent applications as radio, television, and aircraft navigation systems to operate as you cross borders from one country to another.

Because it's desirable to allow certain types of equipment such as portable telephones, microwave ovens, and similar devices to operate without their manufacturers and owners having to acquire licenses, certain frequency bands are not regulated. Such frequency bands are referred to as unlicensed bands and are supported by most countries. Some unlicensed bands are available for use on a global basis, although slight variances in the frequencies of other bands occur as you move from country to country. The most popular unlicensed frequency bands are referred to as ISM bands, with the initials ISM referring to Industrial, Scientific, and Medical. As we examine the operating frequencies of wireless LANs later in this chapter, we note that they primarily operate in two unlicensed ISM frequency bands.

Table 2.1 provides a list of communication-based applications and the frequency they commonly use. Note that wireless LANs can operate in the United States in five distinct frequency bands. The first band, which occurs in the 900- to 929-MHz range, represents a band allocated by the FCC for personal communications and which was used by manufacturers of first generation, nonstandardized LANs. Two additional frequency bands used by wireless LANs are in the 2.4-GHz and 5-GHz frequencies. The 2.4-GHz ISM frequency band is used by IEEE 802.11b- and 802.11g-compliant devices, and the 5-GHz ISM band is used by equipment that is compatible with the IEEE 802.11a standard.

Although not considered to represent conventional wireless LANs, two additional types of wireless technology that can be used for mesh network connectivity are WiMax and ZigBee. WiMax, an acronym for Worldwide Interoperability for Microwave Devices, is designed to operate in both unlicensed and licensed frequencies ranging from 2 GHz to 66 GHz. In comparison, the ZigBee Alliance, which represents a group of semiconductor manufacturers, technology providers, equipment manufacturers, and end users, was in the process of developing standards for wireless transmission for data rates of 250 kbps at distances up to 200 feet for both star and mesh network topologies. Although Zigbee-compliant products will operate in the 2.4-GHz ISM frequency band (which represents a global unlicensed band), they will also operate in two other unlicensed bands. Those bands are 915 MHz for use in North America and 888 MHz for use in Europe.

To assist readers who may lack familiarity with IEEE wireless LAN standards as well as knowledge of WiMax and ZigBee, let's briefly focus our attention on the basic IEEE 802.11 standards, the IEEE 802.16 standard and the relationship of ZigBee to the IEEE 802.15 radio frequency standard.

2.2 IEEE Standards

In the United States the American National Standards Institute (ANSI) delegated responsibility for the development of LAN standards to the IEEE.

Table 2.1   Common Wireless Applications
and Their Frequencies

Application	Frequency Band Used
AM radio	535–1635 kHz
Analog cordless telephone	44–49 MHz
Television	54–58 MHz
FM radio	88–108 MHz
Television	174–216 MHz
Television	470–806 MHz
Wireless data	700–720 MHz
Cellular	806–890 MHz
Digital cordless	900 MHz
Personal communications	900–928 MHz
Nationwide paging	929–932 MHz
Satellite telephone uplink	1610–1625.5 MHz
Personal communications	1850–1990 MHz
Satellite telephone downlink	2.4835–2.5 GHz
IEEE 802.11b/g wireless LANs	2.4 GHz
IEEE 802.15.4/Zigbee alliance	2.4 GHz
IEEE 802.16/WiMax	2–66 GHz
IEEE 802.11a wireless LANs	5 GHz
Large dish satellite TV	4–6 GHz
Small dish satellite TV	11./–12.7 GHz
Wireless cable TV	28–29 GHz

As a result of this delegation the IEEE initially developed standards for wired Ethernet and Token Ring LANs during the 1980s. Approximately 20 years later the IEEE developed its first wireless LAN standard which is referred to as the 802.11 standard.

The 802.11 Standard

The first wireless LAN standard developed by the IEEE was the 802.11 standard. This standard defined the use of three physical layers for wireless communications: infrared, Frequency-Hopping Spread Spectrum (FHSS), and Direct Sequence Spread Spectrum (DSSS). Vendors developed products that used FHSS and DSSS for wireless LANs during the late 1990s, however, to the best of this author's knowledge no products were ever developed to follow the IEEE 802.11 infrared communications standard.

FHSS

Under the frequency-hopping spread spectrum a station transmits for a small period of time at one frequency, with the period referred to as dwell time, and then hops to a different frequency to continue communications. The frequency-hopping algorithm is known to each LAN station, enabling each station to adjust its transmitter or receiver according to its mode of operation. One of the more interesting aspects of FHSS is the fact that its origin dates to the actress Hedy Lamarr who suggested the technique to the United States War Department during WWII as a transmission security mechanism.

DSSS

Direct sequence spread spectrum represents a second transmission technique developed by the military to overcome enemy jamming. Under DSSS a spreading code is applied to each bit to spread the transmission. At the receiver a "majority rule" is applied. That is, if the spreading code is 5 bits and the bits received were 10110, because three bits are set, the receiver would assume the correct bit is a 1. Under the IEEE 802.11 standard an 11-bit spreading code is employed.

Utilization

The initial use of 802.11 wireless LANs was limited due to their relatively low data rate. This was because each of the three physical layers was only defined for operations at 1 Mbps and 2 Mbps. Recognizing the need for a higher data transmission rate resulted in the IEEE initially developing two extensions to the basic 802.11 standard. Those extensions are known as the 802.11a and 802.11b standards.

The 802.11a Standard

The IEEE 802.11a standard defined a series of new modulation methods that enable data transmission rates up to 54 Mbps. The higher data rates are obtained by the use of Orthogonal Frequency Division Multiplexing (OFDM), a technique in which the frequency band is divided into subchannels that are individually modulated. The IEEE 802.11a standard defines operations in the 5-GHz frequency band. This means equipment supporting the standard is not backward-compatible with the basic 802.11 standard because that standard defines operations in the 2.4-GHz frequency band. In addition, because high frequencies attenuate more rapidly than low frequencies, this results in 802.11a wireless LAN stations having a shorter range than stations operating in the 2.4-GHz band. This in turn requires an organization to deploy more access points to obtain a similar geographical area of coverage than would be required via the use of access points operating in the 2.4-GHz band.

The 802.11b Standard

The second extension to the basic IEEE 802.11 standard is the 802.11b standard. Under the IEEE 802.11b standard, DSSS was used with two new modulation methods to provide a data transfer rate of 11 Mbps and 5.5 Mbps. The 802.11b standard also provides compatibility with 802.11 DSSS equipment operating at 2 Mbps or 1 Mbps. To provide this compatibility, the IEEE 802.11b standard specifies the use of the 2.4-GHz frequency band.

The 802.11g Standard

A comparison of the IEEE 802.11a and 802.11b extensions to the 802.11 standard indicates advantages and disadvantages associated with each. Although the 802.11a standard provides a higher data transfer rate, its use of the 5-GHz frequency band results in a shorter transmission distance. Similarly, in a reverse manner, the IEEE 802.11b standard provides a greater transmission distance but lower data rate than obtainable from the use of 802.11a-compatible equipment. By combining the modulation method used in the 802.11a standard with the frequency band employed by the 802.11b standard, the IEEE provided a mechanism to extend both the data rate and transmission range of wireless LANs, resulting in the 802.11g standard. To provide backward compatibility with the large base of 802.11b equipment, the 802.11g standard also supports DSSS operations at 11, 5.5, 2, and 1 Mbps. Thus, the relatively new IEEE 802.11g standard can be considered to represent a dual standard because it provides 802.11b compatibility.

The WiMax Standard

WiMax represents a wide area networking technology that can be used to transmit broadband signals at distances up to approximately 30 miles.

WiMax was standardized by the IEEE as the 802.16 standard, which was published in April 2002.

WiMax has some similarities to wireless LANs; however, it is also significantly different from the IEEE series of 802.11 standards. Concerning similarities, like the 802.11-compliant products WiMax involves the use of client stations using antennae to communicate with a centralized station. That centralized station is referred to as a central radio base station under 802.16 terminology and is designed to provide an alternative to cabled access networks, such as coaxial-based systems operated by your cable company in which you use a cable modem to access the Internet and a Digital Subscriber Line (DSL) for Internet access commonly offered by your local telephone company.

Although the original 802.16 standard was defined for use in the 10- to 66-GHz frequency band that represents spectrum available on a global basis, such high frequencies represent a significant deployment problem. This problem results from the fact that high frequencies have short periods, which restricts transmission to line-of-sight operations. Recognizing this problem, the IEEE developed an extension to the 802.16 standard known as 802.16a. This new standard defines an extension of WiMax to operate at lower frequencies in the 2- to 11-GHz band, to include operations in both licensed and unlicensed frequency bands.

As with the series of wireless LAN standards, WiMax supports point-to-point and point-to-multipoint transmission. In addition, the 802.16a standard extension is designed to add support for mesh networking. Also similar to its wireless LAN cousins, WiMax supports several modulation methods, such as single carrier and Orthogonal Frequency Division Multiplexing (OFDM), with the latter support occurring via the use of 256 and 2048 point transforms. Unlike its wireless LAN cousins, WiMax represents a much more sophisticated transmission method and includes support for Time Division Duplexing (TDD) as well as Forward Error Correction (FEC) employing Reed–Solomon coding.

The ZigBee Standard

In concluding this section describing IEEE standards suitable for wireless mesh networking, we focus our attention upon obtaining a brief overview of a networking technology recently developed to transfer data at distances between 30 to 200 feet while consuming an extremely low amount of power. This technology, which is backed by the ZigBee Alliance, is standardized by the IEEE 802.15.4 specification.

In order to support low-power wireless transmission, ZigBee equipment operates at a relatively low data rate in comparison to the other IEEE standards discussed in this section. ZigBee-compliant products permit data rates of 250 kbps when operating in the 2.4-GHz band where ten channels are available for use. When ZigBee-compliant devices operate in the 915-MHz band where six channels are available for use, their maximum operating rate is reduced to 40 kbps, and operations at 868 MHz where a single channel is available results in a maximum data rate of 20 kbps.

The goal in developing ZigBee technology was to provide a standard for the operation of remote monitoring sensor devices. Such devices typically operate on battery power within an industrial environment. Due to this, batteries powering the remote monitoring and sensor equipment need to last for a year or more to reduce maintenance cost. Thus, the low power requirement of ZigBee-compliant products is especially well suited for remote monitoring and sensor devices. In addition, because ZigBee battery-powered devices can be placed into a sleep mode of operation, this action further reduces the drain on a battery.

In addition to remote monitoring and sensor-based applications in industrial areas, it's expected that ZigBee-compliant products will find their way into home-based applications. Such applications can include wireless home security; remote thermostats for heat pumps, air conditioners, and furnaces; remote lighting; and even remote controls for television and audio systems.

Under the IEEE 802.15.4 standard's addressing scheme, 255 active nodes are supported by a network coordinator. Although most home-based networks will employ a single network coordinator, multiple network coordinators can be linked together to create very large networks. For example, the use of 16 channels in the 2.4-GHz frequency band permits a ZigBee network to contain over 3000 nodes. ZigBee-compliant devices can transmit up to approximately 30 meters in comparison to Bluetooth's 10-meter limit. In addition, a ZigBee network with a single network coordinator can support up to 255 devices in comparison to 8 for a Bluetooth network. Due to these differences, ZigBee represents a more powerful networking technology than Bluetooth, however, it's important to remember that at short distances Bluetooth transmission rates are approximately a magnitude beyond that obtainable by ZigBee products.

2.3 Power Measurements

The ability to understand antenna sensitivity as well as general RF operations depends upon knowledge of power measurements. Thus, in this section we turn our attention to this topic and obtain an appreciation for several terms associated with power measurements.

The Bel

The first power measurement term we discuss traces its roots to the evolution of the telephone network. Named after the inventor of the telephone, the Bel (B) uses logarithms to base 10 to express the ratio of power transmitted to power received. The resulting gain or loss of a circuit is given by the following formula.

$$B = \log_{10}(P_o/P_I)$$

where
   B = power ratio expressed in Bels
   P_o = received or output power
   P_I = transmitted or input power

The rationale for the use of logarithms to base 10 corresponds to the manner by which humans hear sounds. That is, our audio capability perceives sound or loudness on a logarithmic scale. For example, if you estimate, based upon your hearing, that a signal doubled in its loudness, the transmission power actually increased by approximately a factor of ten.

A second reason for the selection of logarithms for use in power measurements results from the fact that changes to a signal in the form of signal loss due to resistance or signal gain due to the use of an amplifier are additive. Thus, the ability to add and subtract when performing power measurements based on a log scale considerably simplifies computations. For example, a 10-*B* signal that encounters a 5-*B* loss and is then passed through a 20-*B* amplifier results in a signal strength of 10 − 5 + 20 or 25 *B*.

Log Relationships

There are two log relationships worth noting that can be used to simplify power measurement computations. First, you can note that the logarithm to the base 10 ($\log_{10}$) of a number is equivalent to determining how many times 10 is raised to a power to equal the number. For example, $\log_{10}100$ is equivalent to determining how many times 10 is multiplied by itself (raised to a power) to equal 100, with the answer being 2. Similarly, $\log_{10}1000$ is 3, $\log_{10}1000$ is 4, and so on.

In examining the preceding equation we can note that under normal circumstances the output or received power can be expected to be less than the input or transmitted power. When this situation occurs the numerator in the preceding equation (P_o) will be less than the denominator (P_I). To simplify computations when this situation arises we can use a second property associated with the use of logarithms. This second property permits us to easily resolve fractional computations as we merely have to prefix the computation with a negative sign to flip its fraction to a whole number. That is,

$$\log_{10} 1/X = -\log_{10} X$$

Once we prefix the computation with a negative sign and flip numerator and denominator, it becomes relatively simple to compute the log. For example, let's assume that the power received is one-hundredth of the transmitted power. Then, our initial computation of the gain or loss becomes:

$$B = \log_{10} 1/100$$

As previously noted, $\log_{10} 1/x = -\log_{10} x$. Thus, using this relationship we obtain:

$$B = -\log_{10} 100 = 2$$

Note that the computational result is negative, which indicates that a power loss occurred and is precisely what we would expect inasmuch as the received power was a very small fraction of the transmitted power. In comparison, a positive Bel value would indicate a power gain, because the output power would be greater than the input power. Although your initial reaction might be dubious concerning a power gain, you need to remember that a signal flowing through an amplifier could result in this situation occurring. In a wireless LAN environment a client transmitting to another client through an access point has its signal regenerated by the AP. If the AP has a higher power level this situation is equivalent to an amplifier in a wired environment.

Although the Bel was used for many years to categorize the quality of a transmission circuit, it gradually lost favor due to the requirement for a more precise measurement. The use of the decibel (dB) provided industry with the precise measurement it sought and for all practical purposes has replaced the use of the Bel. Thus, let's turn our attention to the decibel, which is better known by its abbreviation, dB.

The Decibel

The decibel represents the standard used today to denote power gains and losses. As previously noted, the decibel represents a more precise measurement than the Bel. This is because the dB represents one-tenth of a Bel. To indicate this we multiply the previously noted computation of the Bel by 10 to obtain the computation for the decibel. That is, the power measurement in terms of decibels is computed as follows.

$$dB = 10*\log_{10}(P_o/P_I)$$

where
  dB = power ratio in decibels
  P_o = output power or received power
  P_I = input power or transmitted power

To illustrate an example of the computation of a power ratio in dBs let's return to our prior power ratio computational example. In that example the received power was measured to be one-hundredth of the transmitted power. Thus, the power ratio in decibels becomes:

$$dB = 10 \log_{10} 1/100/1 = 10 \log_{10} 1/100$$

Because $\log_{10} 1/x = -\log_{10} X$, we obtain:

$$dB = -10 \log_{10} 100 = -20$$

In comparing the results of our computations for the Bel and decibel for the same input and output power measurements, note that the decibel is precisely 10 times the value computed for the Bel. Thus, the dB provides the ability for more precise power measurements and today is the preferred power measurement in use.

Decibel above 1 mW

The terms Bel and decibel represent a ratio or comparison between two values, such as input and output power. Although they are important tools, they are not useful for comparing two circuits inasmuch as they do not specify a common input power level. Thus, for comparing two or more circuits we would want to inject the same amount of power into each circuit and observe the level of received power. In the wonderful world of telecommunications testing a 1-mW signal occurring at a frequency of 800 Hz is used in North America. To ensure that you do not forget that testing occurred with respect to a fixed 1-mW signal the term "decibel-milliwatt" (dBm) is used.

Output power with respect to a 1-mW test tone is computed in dBm as follows.

$$dBm = 10 \log_{10} (\text{output power/1 mW input power})$$

Table 2.2   Relationship of Watts and dBm

Power in Watts	Power in Decibel-Milliwatts
.001 mW	−30
.01 mW	−20
1 mW	0
1 W	30
1 kW	60
1 MW	90

Note we use the term dBm to remind us that the output power measurement occurred with respect to a 1-mW test tone. Although the term "decibel-milliwatt" is used in most literature, in actuality dBm means decibel "above" 1 mW because the output or received signal is based upon the input of a 1-mW signal. Thus, 10 dBm more correctly represents a signal 10 dB above or bigger than 1 mW, 20 dBm represents a signal 20 dB above 1 mW, and so on.

One interesting power relationship concerns a 30-dBm signal. A 30-dBm signal is 30 dB or 1000 times larger than a 1-mW signal. Thus, 30 dBm is equal to 1 W. We can use this relationship of 30 dBm being equal to 1 mW to construct a watts to decibel-milliwatt conversion table which is shown in Table 2.2.

Most of the entries in Table 2.2 should be self-explanatory, however, let's review them to ensure we are all on the same path. Let's start our review with the third line in the table, where 1 mW is shown as equivalent to 0 dBm. Because

$$dBm = 10 \log_{10}(P_o/P_i)$$

the only way to achieve 0 dBm is for P_o to equal P_i. Thus, 0 dBm must have an output power of 1 mW. Once we understand that 0 dBm is equivalent to 1 mW, then the other entries in the table are easy to understand. For example, 1 W is 1000 times greater than 1 mW. Because 30 dBm represents a signal 1000 times that of a 1-mW signal, then 1 W is equal to 30 dBm. Similarly, 1 kW is 1000 times greater than 1 W and 1 MW is 1000 greater than 1 kW. Thus, we need to add 30 dBm for each, resulting in 1 kW being equal to 60 dBm and 1 MW being equal to 90 dBm. The only remaining entries to review are the first two. Because .001 mW is one-thousandth of the input power of 1 mW, we obtain:

$$dBm = 10 \log_{10}1/1000/1 = 10 \log_{10}1/1000$$

or

$$\text{dBm} = -10 \ \log_{10}1000 = -30$$

Thus, −30 dBm is equal to .001 mW of power. Similarly, .01 mW requires an output one-hundredth of 1 mW, which is equal to −20 dBm.

The Decibel Isotropic

Because wireless devices communicate via the use of antennae another metric that warrants our attention is decibel isotropic, abbreviated as dBi. This metric is used to define the gain of an antenna relative to a hypothetical antenna that radiates output uniformly in all directions. This uniform radiating antenna only exists in theory and is known as an isotropic antenna. Thus, dBi represents a measurement of how much better an antenna is in comparison to an antenna that transmits signals equally in all directions.

The computation of dBi is based upon the decibel, resulting in the gain (*G*) of an antenna. Specifically,

$$G = 10 \ \log_{10}(I_A/I_i)$$

where *G* is the gain of the antenna in dBi, I_A is the electromagnetic field of intensity measured in microwatts per square meter (mWm²) generated by antenna A, and I_i represents the electromagnetic field of intensity produced by an isotropic antenna and similarly measured in microwatts per square meter, with both measurements occurring at the same distance from the antennae.

Because the dB is commonly used in communications it's often helpful to have access to a table of dB and power ratio values. One such table is provided in this chapter to assist readers in any computations they may have to perform. Table 2.3 contains a decibel reference table, with decibels ranging from 0 dB to 100 dB with the equivalent power ratio for each dB entry. Because dB values are algebraic, you can add or subtract one value from another to obtain a desired value. For example, from Table 2.3 the power ratio for 10 dB is 10, and the power ratio for 20 dB is 100, because doubling the dB increases the power ratio by a factor of ten. Working up from 10 dB, a +3 dB is equivalent to a power ratio of 2 × 10 or 20. Thus, 13 dB is equal to a power ratio of 10 + 20 or 30. Similarly, adding another 3 dB (10 + 3 + 3) for a total of 16 dB results in a power ratio of 50 (10 + 2 × 10 + 2 × 10). If we added another 2 dB for a total of 18, our power ratio would become 65.8 (10 + 2 × 10 + 2 × 10 + 1.58 × 10).

Table 2.3   Decibel Reference Table

dB	Power Ratio
0	1.0
0.5	1.12
1.0	1.26
1.5	1.41
2.0	1.58
3.0	2.00
4.0	2.51
5.0	3.16
6.0	3.98
7.0	5.01
8.0	6.31
9.0	7.94
10	10.00
15	31.6
20	100
25	316
30	1000
40	10000
50	100000
60	1000000
70	10000000
80	100000000
90	1000000000
100	10000000000

Now that we have an appreciation for the use of power measurements let's turn our attention to antennae and more specifically those used in wireless LANs.

2.4 Antenna Systems

An antenna can be considered to represent a conversion device, changing electrical energy into a magnetic force referred to as RF energy and RF energy back into electrical energy. As noted earlier in this chapter the gain of an antenna is measured in dBi. A higher antenna gain is better than a lower value and the overall gain of an antenna indicates its ability to pick up a radio signal.

Antenna Categories

Antennae can be subdivided into two general categories, omnidirectional and unidirectional. An omnidirectional antenna radiates in all directions. In comparison, a unidirectional antenna radiates in only one direction. That direction has a variance, referred to as a beam width, which indicates the directionality of the antenna.

In general, antenna directionality is closely related to antenna gain. By concentrating its RF energy in a specific direction, an antenna obtains more gain. An omnidirectional antenna primarily concentrates its radiated energy into the horizontal plane. Antenna gain is increased by concentrating emitted energy out of the antenna horizontally.

Polarization

As we noted earlier in this section, an antenna converts electrical current into electromagnetic waves that are radiated into space. The direction by which the electromagnetic waves flow is referred to as the polarization or the orientation of the emitted radio waves. In general most antennae radiate either linear or circular waves.

A linear polarized antenna radiates waves entirely in one plane, with the plane representing the direction of wave propagation. In comparison, a circular polarized antenna's plane of polarization rotates in a circle in which one complete revolution occurs during one period of the wave. When the rotation of the wave is clockwise it's referred to as Right-Hand-Circular (RHC), and a counter-clockwise rotation as you might expect is referred to as Left-Hand-Circular (LHC).

Antennae can be horizontally or vertically polarized. When horizontally polarized the majority of the electromagnetic waves flow horizontally or parallel to the Earth's surface. In comparison, a vertically polarized antenna has the majority of its electromagnetic waves flow perpendicular to the Earth's surface.

The electromagnetic radiation pattern is normally shown through the use of two plots. One plot, in the form of a 360° circle, has concentric rings from a center specified in dB as shown in the left portion of Figure 2.2. A second plot indicates the strength of the electromagnetic waves with respect to viewing an antenna from its side. An example of this plot is shown in the right portion of Figure 2.2. Antennae used with most access points are vertically polarized. This results in a radiation pattern that flows in all directions outward from the antenna. Because wireless LANs require a near line-of-sight path due to their low power and high frequency, most access points include dual antennae. Such antennae are referred to as

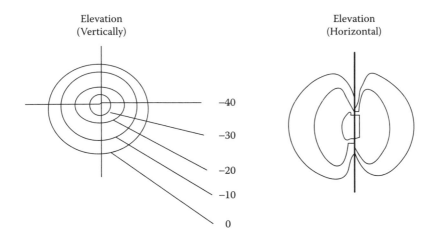

Figure 2.2   Antenna radiation patterns.

diversity antennae and software selects the antenna receiving the better signal as a method to reduce the effect of signal reflections.

Unlike a stereo where you can turn up the power only limited by your speakers to whatever setting you desire, the Federal Communications Commission (FCC) in the United States as well as other regulatory bodies restrict RF transmission power to include that of wireless LANs. That restriction occurs in terms of Estimated Isotropic Radiated Power (EIRP). EIRP considers the transmitter output power, line loss associated with cabling the transmitter to the antenna, and the antenna gain.

Figure 2.3 illustrates the computation of EIRP. Note that gains are added to the Transmit Output Power (TOP) and losses are subtracted to obtain the EIRP. For example, if the transmitter output power is 30 dBm, cable loss is 1 dB, and the antenna gain is 7 dBi, then the EIRP becomes 30 − 1 + 7 or 36 dBm, which just happens to represent the maximum EIRP

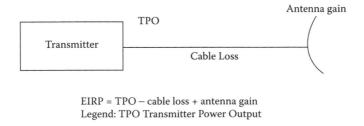

EIRP = TPO − cable loss + antenna gain
Legend: TPO Transmitter Power Output

Figure 2.3   Computing EIRP.

allowed by the FCC. This means that in order to stay within legal limits, you need to consider all three components of EIRP.

Directionality and EIRP

Antennae can obtain more gain by focusing their radiated energy in a specific direction. For example, an omnidirectional antenna primarily concentrates its energy into the horizontal plane in a 360° radius because transmitting RF energy vertically is not normally beneficial. In comparison, a unidirectional antenna concentrates its energy in a narrow beam, with the term "beam width" used to refer to the angular width in degrees between the half power points (3 dB down from the maximum) of the major lobe in either the elevation or azimuth radiation pattern.

EIRP represents the effective power in the main lobe of a transmitter relative to an isotropic radiator that has a 0-dB gain. EIRP equals the sum of the antenna gain in dBi and the power injected into the antenna less cabling loss. For example, assume your transmitter outputs a 16-dBm signal via a cable with a 1-dB loss into an antenna that has a 12-dBi gain. Then, EIRP = 12 dBi + 16 dBm – 1 dBm = 27 dBm. From the preceding, note that a 27-dBm signal represents a signal 500 times above 1 mW. Thus, the effect results in the antenna radiating 500 mW of power. However, because EIRP is less than 36, you can consider either boosting transmission power or using an antenna with a higher gain to enhance transmission and reception of data.

Types of Antennae

There are numerous types of antennae in use today. Some of the more popular types you will encounter include the whip, dipole, loop, patch, and yagi. Figure 2.4 illustrates each of the previously mentioned types of antennae.

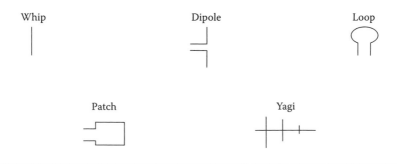

Figure 2.4   Common types of antennae.

Whip Antenna

A whip antenna is a small vertical rod. Although commonly used in automobiles they are not normally used in wireless LANs. This is because the rod does not permit energy to be focused and as a result the whip antenna does not have a high gain.

Dipole Antenna

The dipole antenna represents a simple type of antenna that is very popular in communications applications. The dipole antenna has a straight electrical conductor that measures one-half wavelength from end to end, connected at the center to an RF feed line or cable.

The dipole antenna is a balanced antenna due to its symmetry. Because of its design it is also referred to as a doublet. Dipole antennae are omnidirectional and have a relatively low gain with the RF field primarily focused in the vertical direction. Because it's impractical for many communications devices to have both positive and negative dipole areas when a device sits flat on a desk or other surface, the half dipole is used in place of a dipole antenna.

Yagi Antenna

A yagi antenna can be viewed as a dipole with directors and reflectors added before and after the dipole. The reflectors and directors are metal rods of similar length to the dipole that concentrate RF energy into a beam, increasing the gain of the antenna.

A yagi antenna commonly has one reflector and one or more directors, with the overall rating of the antenna being a function of its total number of elements. A yagi antenna can be considered to represent a directional form of a dipole antenna.

Considering Power Limits

As previously noted, FCC regulations restrict transmission to 36 dBm (4 W) EIRP in the 2.4-GHz frequency band. Because transmitter power and antenna gains are cumulative you need to consider both to stay within legal limits. Table 2.4 provides a summary of the relationship of power injected into an antenna, antenna gain in dBi, and EIRP in dB.

Antenna Selection

When the distance between wireless LAN stations increases you can either add repeaters or use directional antennae to better support communications.

Table 2.4 Legal Relationship among Power Injected into an Antenna, Antenna Gain, and EIRP in the 2.4-GHz Band

Power at Antenna (dBm/Watts)	Antenna Gain (dBi)	EIRP (dBm)
30 (1 W)	6	36
27 (500 mW)	9	36
24 (250 mW)	12	36
21 (125 mW)	15	36
18 (62.5 mW)	18	36
15 (31.25 mW)	21	36
12 (15.125 mW)	24	36

For example, a single-element antenna may provide a gain of 6 dBi, whereas a specialized parabolic antenna could provide a gain well over 24 dBi. However, because the EIRP maximum is fixed at 36 dBm, this means you may need to lower the transmit power when using certain types of high-gain antennae. For example, if transmit power is 27 dBm (500 mW) you can only use an antenna with a gain of 9 dBi or less, as any gain over 9 dBi would result in an EIRP greater than 36. Of course, you could lower the level of transmit power, but doing so would defeat the purpose of acquiring a high-gain antenna.

If you have low-power devices you can consider the use of an antenna with multiple elements referred to as an array antenna. Figure 2.5 illustrates a four-element array antenna, which can be used to enhance transmission distance by generating RF energy in a particular direction. Such antennae are referred to as tuned element array antennae. By using different frequencies for each element and changing their transmission phase the

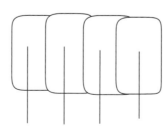

Figure 2.5 A four-element array antenna.

antenna becomes a phased-array antenna, with similar operating characteristics to distant early warning phased array radar. That is, such antennae become highly directional and their use in wireless LANs may occur by the time you read this book.

Receiver Sensitivity

In concluding this section let's turn our attention to receiver sensitivity. The sensitivity of an antenna has a considerable bearing on its ability to receive a signal. Although the FCC does not place limits on receiver sensitivity, IEEE standards denote receiver performance standards. For operations in the 2.4-GHz band, IEEE wireless equipment should have antenna sensitivity less than or equal to –80 dBm. Because 60 dB is equivalent to a power ratio of a million to 1, –80 dBm means that the antenna should be capable of picking up one-hundredth-millionth of a signal.

Wireless LANs are essentially line-of-sight transmission systems. Thus, as you might expect, their transmission distance is greater outdoors than indoors. Table 2.5 lists the transmission distances for a Linksys (now a part of Cisco) 802.11b access point. By appropriately placing repeaters it's possible to extend transmission distances, which enables wireless devices to form a mesh network that spans a considerable distance. As we continue our journey of exploration, we note how homes within communities can be connected to one another in a mesh network environment that enables one or a few Internet connections to be shared among all users instead of requiring each happy homeowner to pay a monthly bill.

Table 2.5   Linksys Access Point Transmission Distances

Data Rate	Indoors (m/ft)	Outdoors (ft)
11	50 (164)	250 (820)
5.5	80 (262)	350 (1148)
2	120 (393)	400 (1312)
1	150 (492)	500 (1640)

Chapter 3

Mesh Network Components

In previous chapters we became acquainted with wireless mesh networking and the radio frequency spectrum primarily available for use by evolving mesh networks. In this chapter we build upon the preceding chapters by turning our attention to common components that function as building blocks for the construction of mesh networks. Such components include what some vendors refer to as "wireless modems," mesh network bridges, routers and extenders, access points, and switches. After examining a series of components as separate entities, we focus on the integration of components to form a wireless mesh network.

3.1 Understanding Mesh Network Components

In this section we examine several mesh network components, commencing our examination with the familiar wireless LAN card which is also referred to as a wireless modem. In doing so we note that many wireless mesh networking components can be used either as is with special software for a limited distance or can be obtained with a different radio technology that provides an extended transmission capability. Thus, the latter provides you with the ability to construct a wireless mesh network over a greater geographical area with a minimum number of access points or routers required for servicing clients.

Wireless LAN Cards

With special software that turns a wireless LAN client into a mesh network participant you can use standardized IEEE 802.11 client hardware. Such hardware is available in many form factors, ranging in scope from PCMCIA adapter cards inserted into a laptop or notebook card slot to PCI adapter cards that are inserted into a desktop computer's system expansion slot and USB combined memory modules and wireless LAN technology. By adding software that provides routing and relay capabilities, the conventional wireless LAN client becomes a participant in a wireless mesh network. Thus, it becomes possible for client computers within a building or in a neighborhood where homes are in close proximity to one another to share one or a few high-speed Internet connections.

When it's not physically possible to communicate within a geographic area using conventional wireless LAN client technology you can consider the use of RF technology developed to extend the transmission distance of client stations while operating in one of the unlicensed ISM frequency bands. One example of this type of product is the 6300 Wireless Modem Card marketed by MeshNetworks of Maitland, Florida.

MeshNetworks 6300 Wireless Modem Card

Through the use of a proprietary Quadrature Division Multiple Access (QDMA) radio protocol that uses the IEEE Direct Sequence Spread Spectrum (DSSS) transmission technique in the ISM 2.4-GHz frequency band, MeshNetworks extended the transmission range of client stations. This transmission extension enables communications of approximately one mile for clear line of sight in comparison to an IEEE 802.11b wireless LAN's maximum range of approximately 300 feet. The QDMA protocol used by MeshNetworks in their wireless modem card includes an enhanced error correction capability over IEEE 802.11 technology as well as a real-time equalization algorithm. The latter uses training signals to automatically adjust a receiver to a transmitter, enabling a receiver to compensate for a rapidly varying signal commonly generated in a mobile environment where signal strength can increase or decrease based upon the movements of two devices communicating on a peer-to-peer basis. Due to the overhead associated with the use of enhanced error correction and real-time equalization, the overall throughput of the MeshNetworks 6300 wireless modem card is approximately 6 Mbps in burst mode in comparison to an IEEE 802.11b wireless LAN's maximum data rate of 11 Mbps.

Figure 3.1 illustrates the MeshNetworks 6300 wireless modem card. This card is fabricated as a Type 2 PCMCIA card that supports data rates

Figure 3.1   The MeshNetworks 6300 wireless modem card. (Photograph courtesy of MeshNetworks)

of 1.5, 3, and 6 Mbps. The card's output power is up to 22 dBm in the 2.4-GHz frequency band using the company's proprietary QDMA protocol.

The QDMA wireless technology used by MeshNetworks uses four separate, nonoverlapping channels in comparison to the three channels used by IEEE 802.11b/g equipment. One channel is used for control purposes and the remaining three channels are used for data transfers. The control channel functions as a coordinator for the transmission of data between devices within a given geographic area.

Figure 3.2 illustrates the channel relationships between MeshNetworks QDMA protocol and equipment conforming to the IEEE 802.11b/g standards. Although at first glance it may appear that the two protocols can interfere with each other, it's important to note that IEEE 802.11b/g devices are limited to operating within one of three available channels. Because the multi-channel QDMA protocol can use any data channel for transmission via a dynamic channel allocation capability, it becomes possible to select a data channel that minimizes interference with an existing IEEE 802.11b/g wireless LAN. Of course, if there were three IEEE 802.11b/g wireless LANs within close proximity of one another each operating on a different channel frequency, the MeshNetworks QDMA protocol's dynamic channel allocation process would more than likely experience difficulty in selecting a channel with minimal interference.

One of the more interesting aspects of MeshNetwork's wireless products is the manner by which they provide a location capability. Modern cell phones now include Global Positioning System (GPS) technology and

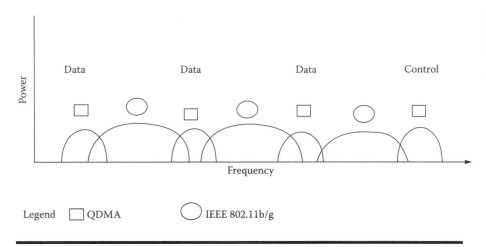

Figure 3.2 QDMA and IEEE 802.11b/g wireless LAN channel relationships.

depend upon satellites to provide 911 telephone location services. In comparison, MeshNetworks includes a location provider mechanism built into its QDMA radio protocol. This location provider enables other radios in the mesh network to use signal triangulation to determine the position of mesh network clients within an accuracy of approximately ten meters. Because the QDMA radio location method does not depend upon the use of GPS satellites, it operates in many locations where it is difficult to obtain a GPS position, such as within buildings or when traveling in a tunnel or through a canyon.

Client Software

As an alternative to the use of the 6300 wireless modem card, MeshNetworks as well as other vendors are marketing software that turns IEEE 802.11 clients into participants on an 802.11 technology-based mesh network. Such software, which was being tested by several vendors when this book was prepared, includes a routing algorithm that enables a conventional wireless network adapter to become a participant in a mesh network in which all other participants use the same wireless RF technology. Although such clients cannot communicate with proprietary RF technology, such as the MeshNetworks 6300 wireless modem card, due to the use of different RF technologies, higher data transfer rates become available, although transmission distance is considerably reduced. This is because IEEE 802.11b/g comparable equipment has a maximum data transfer rate of 11 Mbps (802.11b) or 54 Mbps (802.11g) in comparison to the 6300 wireless modem cards 6-Mbps data rate, whereas the transmission rate of approximately 300 feet for IEEE comparable devices pales in comparison to the approximately one mile range of the 630 modem card. Although there is a tradeoff between the data transfer rate and transmission distance, it's important to note that the conversion of a standard wireless LAN into a mesh network environment can greatly extend the range of the network within a building or on a campus. If the building has a large atrium it may not be possible to position access points within the atrium and run cables to a wired network. However, by using wireless clients as relay devices, you can extend your wireless infrastructure within the building.

Wireless Mesh Router

In a conventional wired environment a router functions as a relay device, routing traffic from one network destined to another toward its final destination. In a wireless networking environment a router performs a similar function, however, routing represents one of many functions performed by this device. Other key functions include the use of a mobile routing algorithm enhanced by an error-correction algorithm to minimize the effect of RF errors, an extended over-the-air transmission range in comparison to IEEE 802.11 LAN devices, and, perhaps most obvious, the ability to support mesh networking. In this section we briefly look at two products, one from MeshNetworks and a second from Nova Engineering.

MeshNetworks Wireless Router

Similar to its client cards, MeshNetworks wireless routers operate in the 2.4-GHz ISM frequency band using the vendors' proprietary QDMA protocol and have an output power of up to 22 dBm. This wireless router functions as a relay or a range extender, enabling clients distant from the vendor's Intelligent Access Point (IAP) to gain access to that device.

Figure 3.3 illustrates the MeshNetworks wireless router shown mounted on an outdoor utility pole. This device can be installed either indoors or outdoors and measures 6.25 inches × 6.25 inches × 4 inches without an antenna. With a weight slightly over three pounds, it's a relatively small and lightweight device that can be used both in-building and as a geographical area extender. This router uses an omnidirectional 7.5-dBi antenna and like the vendor's client 6300 modem, can operate at 1.5, 3, and 6 Mbps.

The MeshNetworks wireless router supports industry standard IP protocols while adding a proprietary "hopping" technology that enables client traffic to be moved around local congestion areas. Because the router

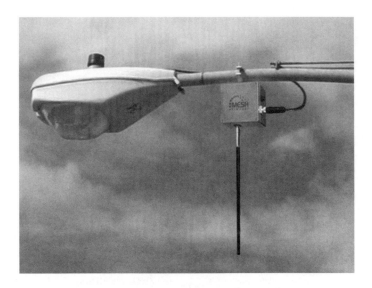

Figure 3.3   The MeshNetworks wireless router shown mounted on an outdoor utility pole. (Photograph courtesy of MeshNetworks)

represents a fixed device, it also functions as a fixed reference point for the triangulation of clients.

Nova Engineering NovaRoam 900 Router

Unlike MeshNetworks products that operate in the 2.4-GHz frequency band, Nova Engineering selected the 900-MHz ISM band for its products. Thus, from a frequency perspective MeshNetworks and Nova Engineering wireless mesh networking products are incompatible with each other.

Another key difference between the MeshNetworks and Nova Engineering routers concerns their utilization. Unlike MeshNetworks routers that are designed to support wireless clients, the Nova Engineering NovaRoam 900 router provides a 10/100-Mbps Ethernet port that enables clients on wired LANs to communicate through the ad hoc mesh networking capability of the router.

The NovaRoam 900 router supports transmission distances that vary with the burst data rate setting of the device. For example, at a data rate of 159 kbps the estimated maximum line-of-sight range is approximately ten miles, whereas at a maximum data rate of 1.008 Mbps the estimated range is three miles. Similarly, power output varies from 800 mW at 159 kbps to 590 mW at 1.008 Mbps. When expressed in dBm the power output varies from 29 dBm at 159 Kbps to 27.7 dBm at a data rate of 1.008 Mbps.

Figure 3.4   The NovaRoam ED900 wireless router uses a conventional Ethernet port to enable wired LAN clients to participate in a wireless mesh network. (Photograph courtesy of Nova Engineering)

Overview

At the time this book was developed Nova Engineering marketed several versions of its NovaRoam 900 router. Figure 3.4 illustrates the NovaRoam ED900 which can be viewed as a successor to the vendor's original NovaRoam900. The ED900 operates in the 915 MHz ISM band (902–928 MHz) using DSSS transmission. Either bridge or router operations can be supported by each ED900. Thus, if your organization wanted to intercon- nect two geographically separated locations, you could configure each ED900 to operate as a bridge, alleviating the necessity to configure stations on the wired network connected to the router with network addresses. However, to obtain the ad hoc networking capability of the ED900, you would use the default routing capability of each router. In doing so, each wired network connected to the router would have a unique subnet and the wireless port on each router would be on the same network.

Through the use of a high-gain antenna, transmission distances up to 22 miles may be possible, however, when operating in the ISM band a maximum EIRP of 36 dB may limit the gain of a selected antenna due to FCC regulations. Because Nova Engineering's routers are also used in many military applications including Unmanned Aerial Vehicles (UAV) the vendor also supports operations in a different frequency band for such applications. Both DSSS and Frequency Hopping Spread Spectrum (FHSS) transmission are supported for Nova Roam 900 operations and customer variances can include a change in the hopping set which makes military applications more difficult to jam.

Operation

Each NovaRoam 900 must participate in a unique Ethernet network. To configure a NovaRoam 900, you can use a terminal emulation program operating on a PC which in turn is directly cabled to the router's auxiliary port. After entering the default password of "novaroam," which should obviously be changed, you can configure the router to include its Ethernet and wireless interfaces addresses, operating channel, data rate, and its routing table.

The Ethernet interface settings entry permits you to assign an IP address to the Ethernet interface. Similarly, the radio settings menu selection provides you with the ability to enter an IP address that will be assigned to the wireless interface. A third required setting involves selecting the channel in the 902- to 928-MHz frequency band the router will use. That channel must be the same as the channel used by other NovaRoam 900s in your network. Because all NovaRoam 900 routers in your network must operate at the same data rate, a fourth setting permits you to specify the data rate used for communications. Once you configure the IP addresses for the wired and wireless interfaces, the RF channel, and data rate, you can add routes into the routers' route table that define how routers communicate with each other.

Route Table Utilization

The purpose of the NovaRoam 900's route table is to define routes from the router being configured to other networks. The addition of a route into the NovaRoam 900's route table requires the entry of four metrics. Those metrics include the destination IP address of the route, its subnet mask, the gateway IP address, and the number of hops from the router to the destination network. Concerning the latter, Nova Engineering refers to the hop count as a "metric".

To illustrate the configuration of route table entries let's assume you acquired two NovaRoam 900s you intend to use as nodes in your network. Let's further assume that the Ethernet and RF IP addresses configured for the first router are 10.1.1.1 and 10.20.20.1, respectively. Let's further assume that the Ethernet and RF IP addresses for the second router are 10.2.1.1 and 10.10.10.2, respectively. Thus, the two routers and their Ethernet and RF IP addresses would appear as shown in Figure 3.5.

In examining Figure 3.5 note that the IP address block 10.0.0.0 through 10.255.255.255 represents a Class A RFC 1918 group of addresses. As a refresher, such addresses represent IP addresses that are used for private Internets and therefore do not represent unique Internet addresses. If we assume that each router will be considered as part of an overall RF network then it becomes apparent why the RF IP addresses for each node are on the same IP network.

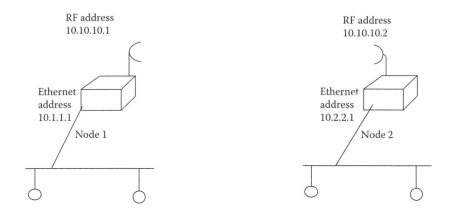

Figure 3.5 Assigning RFC 1918 addresses to the Ethernet and RF interfaces of two NovaRoam routers.

Let's assume that we need to provide a route table entry to allow node 1 to communicate with node 2. As a refresher, the NovaRoam 900 uses a route statement whose format is shown below.

```
Destination IP Address Subnet Mask Gateway, IP Address
Hops
```

To enable node 1 to communicate with node 2, node 1 would use the IP address of node 2's wireless interface as its gateway. Thus, the route table entry added to node 1 would be as follows.

```
10.2.1.0 255.255.255.0 10.10.10.2 1
```

The above entry informs the router that any traffic destined for the 10.2.1.0 network must use the router whose RF IP address is 10.10.10.2. Similarly, you would enter the following route table entry into node 2 to allow it to communicate with node 1.

```
10.1.1.0 255.255.255.0 10.10.10.1 1
```

Applications

Previously in this section we noted that the NovaRoam 900 is used in military applications to include pilotless UAVs. In the commercial sector, one of the more interesting applications of the NovaRoam 900 wireless router is its capability to provide Internet access over an extended geographic area. According to Nova Engineering, the use of three NovaRoam 900 routers enabled wireless Internet access at a distance of 15 miles from a base station router that was connected to the Internet. To obtain this

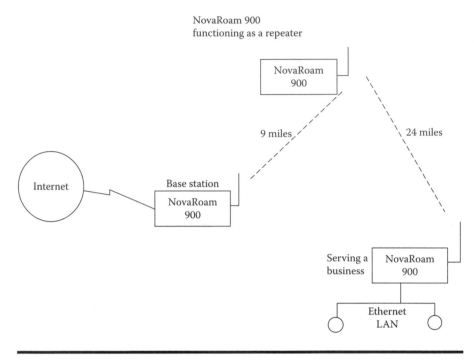

Figure 3.6 Using the Nova Engineering NovaRoam 900 to provide long distance wireless connectivity to the Internet.

extended transmission distance, a base station was located at an elevation of 320 feet. A second NovaRoam router was placed approximately 9 miles away. This router was configured to operate in repeater mode and was placed on a peak at an elevation of 2760 feet. A third NovaRoam 900 was located at a business that required high-speed Internet access but due to its location was unable to acquire the required service. This business, however, was within the line of sight of the NovaRoam router functioning as a repeater and positioned on the mountain peak. Figure 3.6 illustrates the relationship between the three NovaRoam routers.

Each of the routers shown in Figure 3.6 was configured with a yagi antenna that provided 11 dBi of gain. The antennae were horizontally polarized as a mechanism to minimize interference from any adjacent wireless services. Each yagi antenna has a horizontal beamwidth of approximately 40°, however, due to unequal path lengths, the router functioning as a repeater has its beam aimed toward the router serving the business instead of toward the base station.

Access Point

In a mesh networking environment an access point provides a similar level of functionality to that of an AP used in a conventional wireless LAN. That is, the AP operates as a bridge, providing a mechanism to forward data from the RF world's wireless network to the wired network and vice versa.

Operation

A wireless mesh network access point operates at the Media Access Control (MAC) layer. It uses MAC addresses as a decision criterion to forward, flood, or filter data between the wireless and wired networks. To accomplish this task the AP learns MAC addresses associated with its wireless and wired sides, which are considered to represent distinct ports that enable the AP to construct port/address tables.

When the AP is powered on, its port/address table is empty. Thus, the first frame of data received is flooded to the port other than the port on which it was received. At the same time the access point observes the source MAC address and the port that the frame was received on, entering that address and port into its port/address table. If a subsequent received frame has the same destination address as an entry in the APs port/address table, the frame is forwarded to the port associated with the destination address. The only exception to this situation is when the port the frame was received on matches the port in the port/address table associated with the destination address. Because it would be illogical to regenerate the frame back onto the port it was received on, the frame is filtered or discarded. Thus, an access point operates based upon the three Fs: forwarding, filtering, and flooding.

Types of Access Points

An access point can become a participant in a wireless mesh network through the addition of software to an IEEE 802.11-compatible AP. As an alternative, an AP can be fabricated to employ a proprietary RF transmission method and use specialized software. Typically, the use of a proprietary RF method allows extended transmission distances beyond the typical range of IEEE 802.11 access points. One example of a proprietary access point used in a wireless mesh network is the MeshNetworks Intelligent Access Point 6300 (IAP6300). Thus, let's turn our attention to this product.

Figure 3.7 MeshNetworks IAP 6300 can be mounted indoors or outdoors. (Photograph courtesy of MeshNetworks)

MeshNetworks IAP 6300

MeshNetworks IAP 6300 is similar to the vendor's previously described router in that it can be used either indoors or outdoors and supports its proprietary QDMA RF modulation protocol. The use of this protocol enables a burst data transmission rate up to 6 Mbps similar to the vendor's wireless modem card. Figure 3.7 illustrates the IAP 6300 mounted in an outdoor environment. IAPs like the vendor's routers can be mounted on utility poles, buildings, billboards, and other locations that provide a relatively unobstructed level of transmission.

Because a mesh network enables clients to relay messages, the location of the IAP is not critical. That is, as long as clients can communicate with other clients and one client can communicate with an IAP, then all clients obtain this capability. Due to the routing algorithm built into MeshNetworks' products the use of IAPs becomes one of traffic support instead of focusing upon the geographic area of coverage. Thus, you would monitor traffic in hot spots where wireless coverage is provided to determine if an increase in capacity is required. If so, you can add additional IAP 6300s without having to worry about their configuration because the IAPs operate at layer 2 in the protocol stack. This means the new IAPs can automatically integrate themselves into the existing network.

In addition to acting as a bridge between your wireless and wired networks, the IAP provides a fixed reference point for the location of mobile devices. In conjunction with another IAP or the vendor's router a mobile client can be triangulated so that its position becomes known. In addition, software used by IAPs communicates with the vendor's Mobile Internet Switching Controller (MiSC) as well as the vendor's routers and client devices. When communicating with the MiSC, the IAP becomes capable of observing the health of the entire network.

Switching Controller

Although many vendors offer client cards, routers, and access points, some vendors include a switching controller in their product line. You can view a switching controller as a centralized hub of a mesh network as it can be interconnected via wired facilities to distributed access points. The switching controller typically includes a series of management functions including monitoring network events and providing detailed and summary statistics concerning the operation of the network. Other vendor products may include additional functionality, such as operating as a bridge between the wireless mesh network and a wired network. To obtain an appreciation of the functionality of a switching controller we focus our attention upon the MiSC marketed by MeshNetworks.

MeshNetworks MiSC

MeshNetworks' mobile Internet switching controller represents a scalable computer that provides routing, switching, and management functions to a mesh network. MiSC functionality includes operating as a DHCP/DNS server for clients connected to the device via IAPs. That connection occurs through the use of leased lines or cables routed through a building or campus area. Because the MiSC also provides authentication of clients as well as being capable of supporting IP VPN encryption, it performs several security-related functions.

In a mesh-enabled architecture the MiSC is interfaced to one or more IAPs, providing centralized control of the wireless network. Due to its modular design, the MiSC can provide access to the Public Switched Telephone Network (PSTN) and the Internet for clients reaching the MiSC via one or more IAPs. Now that we have an appreciation for the major components used to construct a wireless mesh network, let's examine how these components interact to form this type of network.

3.2 Integrating Components

Because most wireless mesh networks resemble the structure of wireless LANs, it would come as no surprise that the interrelationship of mesh network components is very similar to that of a wireless LAN. In fact, because some vendors market software that literally rides on top of wireless LAN components to create a mesh network, such components from a hardware perspective are not distinguishable from wireless LAN components.

To obtain a general appreciation for the manner by which wireless mesh network components are used, let's turn our attention to Figure 3.8. That figure illustrates the relationship between wireless LAN cards (inserted into laptop or notebook computers) moving into and out of an ad hoc network, routers used as network extenders, and two access points, with the latter used to provide mesh network clients access to the Internet via a common switching controller.

In examining the components shown in Figure 3.8 note that the wireless mesh router functions as an extender or bridge, enabling a client that would otherwise be outside the range of an access point to communicate

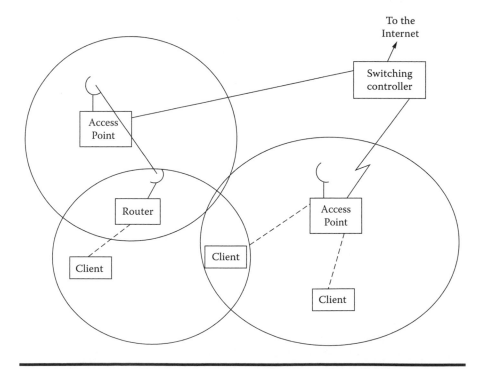

Figure 3.8   Interrelationship of common wireless MeshNetworks hardware components.

with that device. Each of the access points shown in Figure 3.8 is in turn shown connected via wired lines to a common switching controller. That controller can be viewed as functioning similar to a wireless LAN switch connected to limited functioning access points. This is because the switching controller provides support for the issuing of IP addresses and performs other functions that are normally built into access points. By doing so it becomes easier to manage the resulting network.

Chapter 4

Routing Protocols

Similar to other evolving technologies, wireless mesh networking has a number of vendors offering features that are either proprietary or are based upon research that has yet to be standardized. One such area, which is the focus of this chapter, is the routing protocols used by wireless mesh networks.

At the time this book was prepared wireless mesh networks were being standardized by the Internet Engineering Task Force (IETF) as a Mobile Ad hoc NETwork (MANET). Although there are many MANET protocols that have been defined, very few have actually been implemented outside academic publications and research laboratories. One such protocol that shows promise to be widely used is the Ad hoc On-demand Distance Vector (AODV) routing protocol, and a second protocol is known as the Topology Dissemination Based on Reverse-Path Forward (TBRPF) routing protocol, both of which are discussed in this chapter. However, prior to doing so, the first section of this chapter provides a brief overview of MANET, including its characteristics and the goals of the IETF working group established to standardize one or more routing protocols for this mobile network.

4.1 MANET

MANET represents a series of mobile platforms that communicate using wireless transmission and can move in and out of a geographic area in a random or arbitrary manner. In doing so each platform is independent of other platforms because they all have a routing ability, resulting in the network having a self-forming, self-healing, and self-maintaining capability. The resulting ad hoc network can operate in a stand-alone manner or, more likely, will have a connection to the Internet.

Although the ITEF MANET effort is focused upon mobile platforms, it is also applicable for fixed location devices that will more than likely in this author's opinion represent the vast majority of wireless mesh network clients of mesh networks to be constructed during this decade. The rationale for this opinion is based upon the manner by which humans use computers. Although your computer may be transported to a Starbucks, an airport lounge, or a hotel lobby, and use a wireless hot spot to gain access to the Internet, such access commonly occurs with the computer having a fixed location. Thus, although the computer is mobile, its operation typically occurs at a fixed location.

Characteristics

MANETs by their very nature have several key characteristics. Those characteristics include dynamic topologies, a constraint on available bandwidth, some nodes relying upon battery power that constrains available energy, and perhaps most important, limited physical security. In addition, because nodes move in and out of the network randomly, the network needs to be scalable. To obtain an appreciation for each of these constraints let's turn our attention to obtaining additional information about each in the following four sections.

Dynamic Topologies

Network nodes can be mobile computers that are free to randomly move about the network as well as leave the MANET's transmission area of coverage. Their participating in a MANET results in routes from a source node to its destination having a probability of change due to the random structure of the network. Thus, any routing protocol that is used by a MANET should support a dynamic topology.

Bandwidth Constraint

A MANET node transmits information wirelessly. Although wireless transmission has an almost infinite frequency range, those pesky regulators in most countries limit the use of the frequency spectrum to different applications, such as radio, television, cellular phones, and wireless LANs. Because wireless LANs are commonly restricted to the use of 100 MHz or less of frequency and its wired cousin is only restricted to the frequency supported by the media that can range into the GHz band when fiber is used, we can say that MANETs and wireless mesh networks are bandwidth constrained. In addition, the use of wireless transmission results in certain types of transmission impairments that normally do not affect wired transmission. Such impairments can include the movement of vehicles, people, and foliage that causes reflections as well as the effect of sunspots, machinery, and other electrical disturbances. These impairments can result in retransmissions that additionally affect wireless transmissions and can result in periods when the transmission capacity of the network becomes congested. Thus, network managers and LAN administrators must carefully consider the operational requirements of nodes that come together in a random manner to form a wireless mesh network.

Although this effort may not be practical when the mesh network participants join and leave the network in a truly random manner, over a period of time the peaks and valleys in transmission may be recognizable. For example, within a hotel that supports mesh networking the network manager would probably observe transmission peaks in the morning and evening. By examining the use of the access point connected to the Internet during such times, the manager could determine if another access point was required during peak times or if the existing access point was able to satisfy the requirements of hotel guests and staff.

Energy Constrained Operation

Because most MANET nodes are thought of as mobile devices they are considered to be operating on battery power. This means that the ability of the node to operate for a fairly long time will depend upon a variety of design factors including the manner by which it participates in a network. Under the IEEE 802.11 series of wireless LAN standards, nodes can operate in either one of two power modes, active or asleep. When asleep, the node periodically wakes up and listens to the wireless spectrum to see if another device has data destined for the sleeping node. If the node determines that another device has data destined for it, the node in effect continues to operate in its wake-up mode, otherwise, it goes back to sleep. In a MANET similar system designs can be expected to be implemented to conserve energy used to power nodes.

Limited Security

Because MANET uses wireless transmission each node in the network becomes susceptible to such wireless security threats as spoofing, eavesdropping, jamming, and various denial-of-service attacks. In addition, because each node in a MANET has routing capability for other nodes it becomes possible for data from any node to flow through any other node in the network, providing another security threat that normally does not occur in traditional wireless LANs. Thus, although existing wireless LAN security techniques can offer good protection against most threats they do not address the threat of nodes being able to capture data flowing through a node.

Scalability

Because some MANETs can be relatively large with hundreds to thousands of nodes, scalability becomes an issue. Some MANETs, such as mobile military networks could have thousands of nodes within a geographic area and a hotel or airport lounge might have a handful to hundreds of nodes. Although scalability is not unique to MANETs, it must be considered in developing protocols to support routing.

IETF MANET Goals

Based upon the preceding characteristics of mobile ad hoc networks the IETF MANET working group developed several goals to support a peer-to-peer mobile routing capability in a wireless environment. Those goals can be subdivided into near-term and long range and are described in the following two sections.

Near-Term Goals

The primary near-term goal of the IETF MANET working group is to standardize one or more routing protocols to support operations over a mobile network. Such protocol(s) need to provide a "discovery" algorithm that enables newly arriving nodes to learn routes in the network as well as enables remaining nodes to recognize that one or more nodes left the network.

In addition to supporting node discovery MANET protocols need to support connectionless IP service as well as efficiently respond to network changes and traffic variances.

Long-Term Goals

In addition to short-term goals oriented toward developing dynamic routing protocols, the IETF MANET working group has a number of issues that can be considered to represent long-term goals. Those long-term goals include network security, addressing, and the interaction of layer 3 (network) operations with upper and lower layers of the protocol stack. Now that we have a general appreciation for MANET characteristics and the goals of the IETF MANET working group, let's turn our attention to MANET protocols.

4.2 MANET Protocols

Over the past half decade several MANET protocols have been specified in various journals, white papers, and scientific proceedings. Unfortunately, a majority of such protocols have limited implementation data available outside a research environment.

Types of Protocols

Most MANET protocols can be categorized as either being proactive or representing an on-demand or reactive type of protocol. Protocols that can be categorized as being proactive, update the routing information by exchanging route data at periodic intervals. Such exchanged route data is placed into tables in each device and provides information on routing prior to devices requiring route data. Thus, a proactive routing protocol represents a mechanism to reduce network latency inasmuch as there is no need to determine a route when data needs to be transmitted. However, because of the periodic updating of route tables regardless of whether such data is needed, a proactive network routing protocol can have a relatively high overhead.

A second category of MANET routing protocols performs route maintenance only when information needs to flow on a new route. This type of protocol is reactive as it responds to the need to determine a route. Another moniker or name for this type of reactive protocol is "on-demand," because a route is determined only when needed. Because the exchange of routing information occurs when needed, the overhead associated with an on-demand routing protocol is typically less than for a proactive routing protocol. However, because there is no free lunch in communications, the lower overhead occurs at the expense of an increase in latency which results from devices having to learn routes at the time information needs to be transferred. Because of the emergence of the ad hoc on-demand distance vector routing protocol as a popular protocol for use in wireless mesh networks we focus our attention upon how this protocol operates. Once this is accomplished we look at a second protocol. That protocol, which is known as the topology dissemination based on reverse-path forwarding routing protocol represents an experimental, proactive routing protocol currently used by a few wireless mesh networking venders.

The AODV Routing Protocol

As a review, when two or more terminal devices come within close proximity of each other and require the ability to exchange data we have a special type of network. That type of network is referred to as an ad hoc network. This type of network can be formed when students arrive at a classroom; business people or travelers visit a Starbucks coffee shop, a hotel lobby, or an airport lounge; or even when a Federal Emergency Management Agency (FEMA) task group arrives at a site stricken by a flood, earthquake, or another type of natural disaster.

Need for a Routing Protocol

When only two computers are participants in an ad hoc network they can directly exchange information. Thus, there is no need for a routing protocol. However, when three or more computers come together in an ad hoc network a routing capability can become a necessity. To see why this is true consider Figure 4.1 which illustrates an ad hoc network consisting of three wireless hosts or network nodes. If we assume computers A and C are not within range of each other, then the only way they can reach each other is via a route through computer B, because B's service range overlaps A and C. Thus, computer B must learn how to access A and C whereas computers A and C must learn that the route to each other is through computer B. Although routing in an ad hoc network with a larger number of computers becomes more complicated, Figure 4.1 illustrates the need for a protocol that allows computers to use the facilities of other devices to access distant computers.

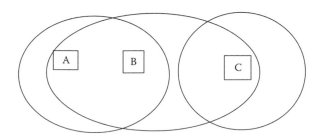

Figure 4.1   An ad hoc network consisting of three wireless computers.

Route Request Message

In an AODV environment routes between computers are created only as needed, hence the term "on-demand" is used to refer to this protocol. Once a route is required to a particular destination, the AODV protocol transmits a Route REQuest (RREQ) message packet that propagates across the wireless network. The format of the RREQ message is shown in Figure 4.2.

In examining the format of the RREQ message packet note the use of destination and origination sequence numbers. The destination sequence number is created by the destination and is included with route information it transmits to requesting nodes. If two routes are defined to a destination the requesting node is required to select the one with the highest sequence number. Here the higher sequence number indicates a newer or fresher route.

In addition to providing a mechanism to select a route when more than one is available, sequence numbers enable AODV to avoid routing loops that could result in messages repeatedly propagating over the same path. To ensure sequence numbers are updated under AODV the broadcast

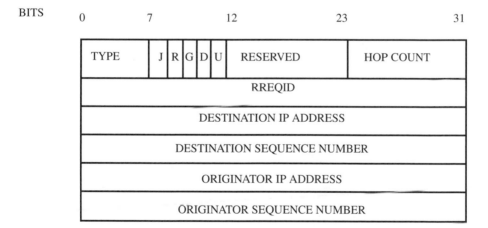

BITS   0   7   12   23   31

TYPE	J	R	G	D	U	RESERVED	HOP COUNT
RREQID							
DESTINATION IP ADDRESS							
DESTINATION SEQUENCE NUMBER							
ORIGINATOR IP ADDRESS							
ORIGINATOR SEQUENCE NUMBER							

Legend

J   Join flag, reserved for multicasting
R   Repair flag, reserved for multicast
G   Gratuitous flag, indicates if a gratuitous RREP should be unicast
    to the node specified in the destination IP address field
D   Destination only flag, indicates only the destination should
    respond to thie RREQ message
U   Unknown sequence number, indicates destination number unknown

Figure 4.2   Route request message format

of RREQ messages as well as RREP (Route REPly) and RERR (Route ERRor) messages which we examine shortly include sequence number fields that are incremented prior to transmission.

Each node that receives a RREQ message packet checks the Destination IP address field to determine if the node represents the destination. If the node is not the destination and does not have a route to the destination it rebroadcasts the RREQ message to its immediate neighbors as well as updates its route table by including a reverse pointer to the originator. This process continues until a route to the destination node is located or the IP datagram transporting the RREQ message reaches its maximum hop count and is discarded.

Route Reply Message

When the RREQ message packet either reaches the destination node or encounters a node with a route to the destination a response is transmitted. That response occurs via the transmission of a route reply message. The RREP message packet whose format is shown in Figure 4.3 flows toward the originating or source node. As the RREP message packet flows through intermediate nodes, such nodes update their route information about source and destination nodes.

In examining Figure 4.3 note that setting the A bit field requires a Route REPly ACKnowledgement (RREP—ACK) message to be returned.

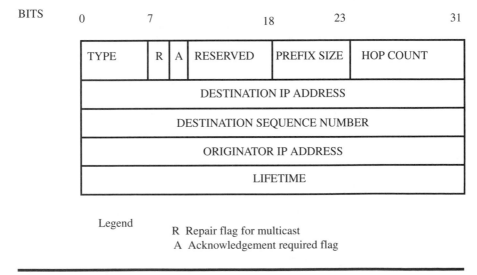

BITS 0   7   18   23   31

TYPE	R	A	RESERVED	PREFIX SIZE	HOP COUNT
DESTINATION IP ADDRESS					
DESTINATION SEQUENCE NUMBER					
ORIGINATOR IP ADDRESS					
LIFETIME					

Legend

R Repair flag for multicast
A Acknowledgement required flag

Figure 4.3   Route reply message format.

Two other fields in the RREP message differ from the RREQ message and also deserve mention. Those fields are the Prefix Size and Lifetime fields.

The 5-bit Prefix Size field when set to a nonzero value specifies that the indicated next hop can be used for any nodes with the same routing prefix as the requested destination. Thus, the Prefix Size field enables a subnet router to provide a route for every host in the subnet defined by the routing prefix. In comparison, the 32-bit lifetime field contains the time expressed in milliseconds for which nodes receiving the RREP message packet consider the route to be valid.

RREQ–RREP Message Flow

To illustrate the flow of RREQ and RREP messages we need a network. Thus, let's assume we have a five-node network as illustrated in Figure 4.4, where node 1 transmits a RREQ message packet in an attempt to determine a route to node 5, the destination computer in the network. Because node 2 represents an intermediate node that is not the final destination this node relays the RREQ message by rebroadcasting it to nodes 3 and 4.

As the RREQ message continues its propagation through the network it reaches node 3. At that node the computer determines it is not the

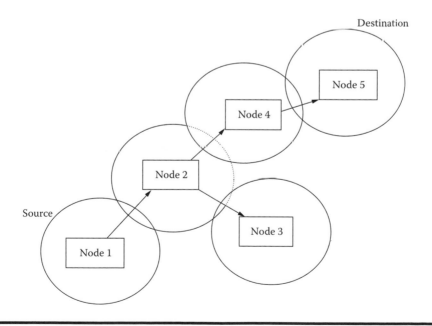

Figure 4.4   RREQ message flow from source to destination in a five-node network.

destination nor does it have a path to another node. Thus, the message packet is dropped by node 3. When the RREQ message packet reaches node 4, that node forwards the packet to node 5, which is the destination node. While the RREQ message propagates through the network each intermediate node observes the value of the packet's originator IP address field, enabling the route information table for the source node to be updated at each node, using the neighbor that propagated the packet as the next hop.

Once the destination node receives the RREQ message it will respond with a RREP message packet. The RREP message packet issued by the destination node, which in our small example is node 5, flows to node 4. At node 4 the computer lookup of its route table indicates that the next hop toward the source node address is node 2. Thus, the RREP is propagated to node 2. At that node another table lookup occurs and node 2 notes that the source node is node 1 and relays the RREP to node 1. This sequence is illustrated in Figure 4.5.

In the preceding example of the determination of a route from source to destination the RREQ message packet was shown flowing to the destination. In actuality, a route can be determined when the RREQ either reaches the destination or an intermediate node that has a route to the destination. That route should be a fresh route, which means that the route is a valid route entry for the destination whose associated sequence

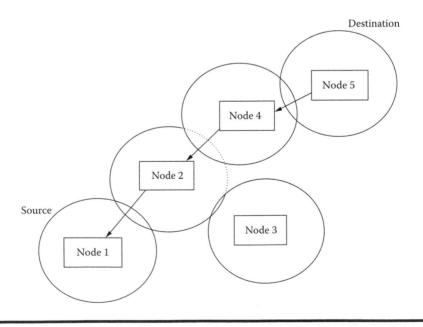

Figure 4.5   RREP message flow in response to a prior RREQ message.

number is at least as great as the sequence number contained in the RREQ message. Now that we have an appreciation for the manner by which routes are determined under AODV let's turn our attention to their deletion.

Route Deletion

Once a route is determined it will remain in each computer's node routing table until the route becomes inactive for a period of time. Each routing table entry contains an active route timeout field which is updated each time the route is used to forward a data packet. If the timeout value expires, the route is then deleted.

A second method can also result in the deletion of a route. That method involves the failure of a node to receive HELLO messages. Thus, let's briefly focus our attention on this message.

HELLO Message

A node can provide connectivity information by broadcasting HELLO messages when it is part of an active route. A HELLO message is a RREP message with a Time To Live (TTL) value of 1 set in the IP header of the message. In the RREP the destination IP address field is set to the node's IP address that is broadcasting the HELLO message. The Destination Sequence Number field is set to the node's latest sequence number, and the Hop Count field is set to a value of 0.

If you examine the format of the route request message previously illustrated in Figure 4.2 you will not see a TTL field. This is because AODV messages are transported via the User Datagram Protocol (UDP). UDP messages are in turn prefixed with an Internet Protocol (IP) header to form an IP datagram, with the IP header containing the TTL field whose value is then set to 1 when an RREP message is used as a HELLO. Figure 4.6 illustrates the formation of an IP datagram that becomes an AODV HELLO message.

If a node fails to hear a predefined number of consecutive HELLO messages from its next hop neighbor, the route table entry will be deleted

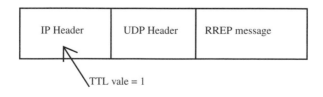

Figure 4.6   An encapsulated AODV HELLO message.

by the listening node. When this action occurs it represents a break in network connectivity which results in the transmission of a RERR message packet. That packet is then propagated back to the source node to inform all nodes that have that route in their routing table that the route is unusable and should be deleted from their routing tables. When a destination becomes unreachable a node will transmit a route error message. Thus, let's turn our attention to the format and operation of this message.

Route Error Message

Figure 4.7 illustrates the format of the route error message. Note that the Dest Count field indicates the number of unreachable destinations included in the message. This field must have a value of at least 1. Similar to other AODV messages, the RERR message is transported via UDP within an IP datagram.

To illustrate the flow of the RERR message let's assume node 5 (previously shown in Figures 4.4 and 4.5) fails or moves outside the range of node 4. When this situation occurs node 4 will not hear HELLO messages from node 5. After a predefined number of HELLO messages are missed node 4 deletes the route to node 5 from its routing table and transmits an RERR message that marks the route to node 5 as invalid. The RERR message is transmitted to neighbor nodes that were using node 4 as the next hop for the route to node 5. Thus, the RERR message is transmitted to node 2. Similarly, node 2 transmits the RERR message to node 1. After receiving the RERR message the computer at each node deletes the route to the unreachable node from its routing table. If a route to the destination that was just deleted becomes required the source node that needs the route will initiate a new route discovery process. Figure 4.8 illustrates the flow of RERR message packets for our 5-node network assuming node 5 became separated from the network.

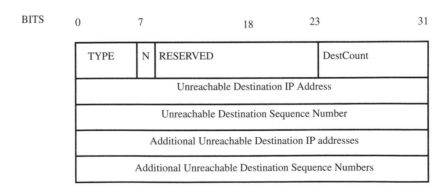

Figure 4.7 Route error message format.

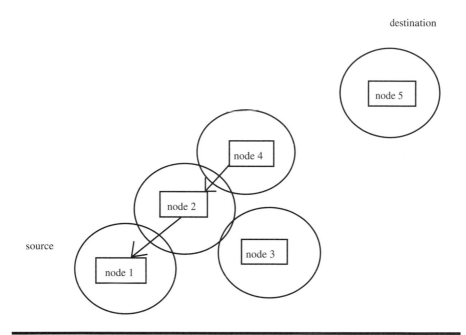

Figure 4.8   RERR message flow assuming node 5 becomes detached from the network.

In concluding our discussion of the AODV routing protocol we turn our attention to a situation referred to as a "gray area." In doing so we note how one vendor modified the AODV protocol to overcome this situation.

Gray Area Considerations

A gray area occurs when the wireless signal between a mobile node and a fixed location becomes too weak for application data to reach its destination. To illustrate how a gray area can occur and the manner by which one vendor modified the AODV routing protocol we need a sample network for illustrative purposes. Thus, let's create one.

Figure 4.9 illustrates two wireless networks interconnected via a pair of wireless bridges. Each network uses the AODV routing protocol to support mobile ad hoc operations even though the bridges interconnecting the two separated geographical areas represent stationary devices. If we assume that just one host is mobile let's examine what happens as it moves from the area served by bridge A to the area served by bridge B.

When the mobile node is within a relatively short distance of other nodes in the network supported by bridge A it can receive a high enough level of signal strength to become a participant in the mesh network at

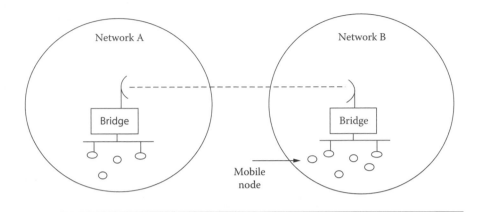

Figure 4.9 The addition of signal strength to control packets can be used to minimize the effect of gray zones.

that location. However, as the mobile node moves from the area serviced by bridge A to the area serviced by bridge B the signal strength between the mobile node and other wireless nodes that make up network A decreases. At a certain distance between networks A and B the mobile node's ability to pick up signals from the wireless nodes in network A degrades to the point where the signal strength is just strong enough for periodic control messages to link the mobile node to network A. In this situation a route from a node in network A to the mobile node may remain in the routing tables of other nodes in network A even though the actual data transfer capability is diminishing toward zero. In fact, as long as one node in network A continues to receive a minimum number of HELLO messages from the mobile node, a route to the mobile node will continue to occur via network A. Only after the mobile node is completely outside the range of other nodes in network A will a node in network B become an intermediate node for the route to the mobile device. The area from which the mobile device loses its ability to communicate with nodes in the network it is leaving until it joins the new network is commonly referred to as a "gray area" or "gray zone."

One interesting method used to counter the gray zone problem involves a change to the manner by which the AODV protocol operates. Under the implementation of the AODV routing protocol by NovaRoam each control packet includes a signal-strength metric. If the control packet signal-strength metric falls below a user-defined threshold an existing link between nodes will be dropped, in effect forcing the affected nodes to find new routes. If we return to the example shown in Figure 4.9, employing the modification to AODV used by NovaRoam, as the mobile node moves toward network B its signal strength to the nearest node in network A diminishes to the point where its connection is dropped. As the mobile node reestablishes a route it moves closer to a node in network B and finds a new route to network B. Thus, the addition of a signal strength indicator can be employed to minimize the gray zone or gray area problem that occurs when truly mobile nodes move between areas where a grouping of nodes forms a wireless mesh network.

TBRPF

The topology dissemination based on the reverse-path forwarding routing protocol is both a mouthful to utter in a short period of time as well as an experimental protocol adapted by a few wireless mesh networking equipment vendors in modified form. TBRPF is defined in RFC 3684 and can be considered to represent a proactive, link-state routing protocol developed for use in mobile ad hoc networks.

Overview

TBRPF is a proactive link-state routing protocol that provides hop-by-hop routing along the shortest paths to each destination in a network. Each node in a network using TBRPF computes a source tree that consists of paths to all reachable nodes based upon practical topological information stored in a topology table in the nodes. To minimize transmission overhead, each node transmits only a portion of its source tree to neighboring nodes, using a combination of periodic and differential updates to keep its neighbors informed of the reported portion of its source tree. Here the differential updates report only changes in the status of neighbors. Each node can optionally report additional topology information with a resulting increase in overhead. In fact, it's possible to configure nodes in a robust mobile network environment using TBRPF as the routing protocol to transmit its full topology to neighbors.

Routing Modules

The TBRPF protocol consists of two main modules. One module is responsible for neighbor discovery and is known as the neighbor discovery module, and the second module performs topology discovery and route computations. The latter is referred to as the routing module.

Neighbor Discovery Module

The TBRPF Neighbor Discovery (TND) module is responsible for discovering neighbors in the network. This protocol enables each node (i) to quickly detect neighbor nodes (j), such that a bidirectional link (I,J) exists between an interface I of node i and an interface J of node j.

Similar to the AODV protocol, the TBRPF neighbor discovery module uses HELLO messages. However, those messages can be differential in that they only report changes in the status of a link. For example, differential HELLO messages would include only the IDs of new neighbors and recently lost neighbors instead of information about all neighbors. This action can result in HELLO messages that are significantly smaller than those of other link-state routing protocols. This enables HELLO messages to be transmitted more frequently and allows faster detection of topology changes that can be an important consideration if nodes are truly mobile.

Because TND is designed to be fully modular and independent of the routing module it only performs direct neighbor sensing. That is, it determines nodes that are one-hop neighbors, resulting in the routing module being responsible for discovering neighbors at a greater distance.

If a node has multiple interfaces, such as a server, TND is separately run on each interface. This action results in the construction and mainte- nance of a neighbor table for each local interface. The neighbor table is responsible for storing the status of each link, such as 1-way, 2-way, or lost. The contents of HELLO messages are then used to update the contents of the neighbor table and the contents of the neighbor table determine the contents of HELLO messages.

The actual specifications for the operation of TND is well thought out as it includes methods to ensure a node will not inadvertently miss the declaration of a link being lost nor will it establish a link that will be short lived. Concerning the former, when a node changes the status of a link it will commonly issue three consecutive HELLO messages. The node at the opposite end of the link will either receive one of the HELLO messages or miss all messages, with the latter situation causing the node to declare the link lost. Concerning short-lived links, nodes must receive a specified number of HELLOs prior to declaring the link to be operational. In this manner the counting of HELLO messages becomes an important criterion for acquiring or losing a link.

Routing Module

Each node operating TBRPF maintains a source tree that indicates the shortest paths to all reachable nodes in the network. Using partial topology information stored in its topology table, each node uses a modified Dijkstra algorithm to compute its topology table. As a refresher, the Dijkstra algo- rithm, which is named after its discoverer, E.W. Dijkstra, solves the problem of locating the shortest path from a point in a graph, referred to as the source, to a specific destination. Because it's possible to find the shortest paths from a given source to all points in a graph at the same time, the problem solved by the Dijkstra algorithm is also referred to as the "single-source shortest paths" problem.

The Reported Subtree

The portion of the source tree that a node reports to its neighbors is referred to as the reported subtree. Thus, if T represents the source tree maintained by each node then RT becomes the reported subtree. Each node reports RT to its neighbors using a periodic topology update, whereas additions and deletions occur as differential updates on a more frequent basis. For example, periodic updates could occur every five or six seconds and differential updates could occur every second.

Through the use of periodic updates new neighbors are informed of the RT, and differential updates are used to rapidly disseminate changes to all nodes affected by the update. The reported subtree (RT) consists of links that include neighbor nodes only when such nodes represent the shortest path to a neighbor. In making this determination a node computes the shortest path from each neighbor to each other neighbor for up to a two-hop distance, using only neighbors as an intermediate node. Although the actual operation of the routing module can be quite complex as the number of nodes in the network increases, we can obtain an appreciation for its basic operation by considering the partial network illustrated in Figure 4.10.

In the example shown in Figure 4.10 let's assume that node 2 selected itself as a parent for all neighbors due to its small ID. As a result, node 2 reports its entire source tree, which in effect is a view of the network without any closed loops. In comparison, nodes 3 and 4 report their one-hop neighbors which represents a small portion of their trees.

TBRPF Packets

TBRPF is a packet-oriented protocol, with each packet consisting of a header, optional header extension, and a body. The latter consists of one or more messages that are filled at the end with padding options that may be necessary for alignment on natural boundaries. The format of the TBRPF packet header is illustrated in Figure 4.11.

In examining the TBRPF packet header note that four bits define the TBRPF version number, with version 4 of the protocol defined when this book was written. The first two flag bits (L,I) specify which header

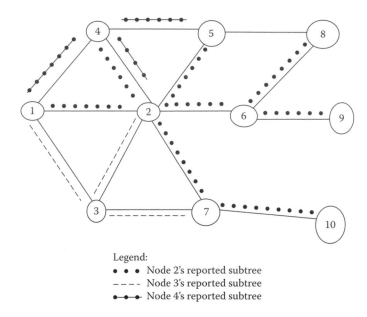

Legend:
• • • Node 2's reported subtree
– – – – Node 3's reported subtree
•—•—• Node 4's reported subtree

Figure 4.10   TBRPF networking example.

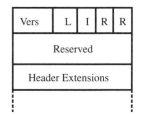

Legend:

Vers   Version
L   Length Included
I   Router ID Included
R   Reserved

Figure 4.11   TBRPF packet header format.

extensions, if any, follow, and the last two flag bits are presently reserved and are set to a value of zero. The L bit when set indicates a 16-bit length field which indicates the length of the TBRPF packet. When the I flag bit is set it indicates that the Router IDentification (RID) field consisting of four bytes is contained in the TBRPF packet.

Packet Body

The TBRPF packet body follows the header. The packet body consists of the concatenation of one or more TBRPF messages and terminates with optional padding to satisfy any alignment requirements.

Messages

Under the TBRPF protocol there are two major categories of messages that flow between nodes. The first category of messages involves three types of HELLO messages that are used for neighbor discovery purposes, and the second category of messages involves topology updates. The basic format of the HELLO message is shown in Figure 4.12.

The 4-bit Type field currently defines three HELLO messages. A Type field value of 2 defines a neighbor request HELLO, and Type field values of 3 and 4 define neighbor reply and neighbor lost HELLO messages, respectively. The 4-bit PRIority (PRI) field indicates the sending node's relay priority, which is expressed as an integer value between 0 and 15. A node with a higher relay priority is more likely to be selected as the next hop than a node with a lower value. The value 0 is reserved for nonrelay nodes, such as nodes that never forward packets originating from other nodes.

A router in normal operation has a relay priority value of 7 and can dynamically change its relay priority value. The Priority field is followed by a 12-bit number that indicates the number of 32-bit IPv4 interface addresses in the message. Each of those addresses represents neighbors of the node.

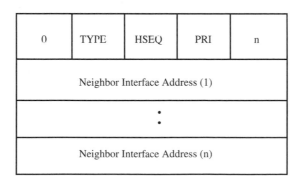

0	TYPE	HSEQ	PRI	n
Neighbor Interface Address (1)				
• • •				
Neighbor Interface Address (n)				

Figure 4.12   TBRPF HELLO message format.

Table 4.1 TBRPF Neighbor Table Entries

Entry	Description
Router ID	The router ID of the node associated with the neighbor interface
Status	The current status of the link (LOST, 1-WAY, or 2-WAY)
Life	Amount of time in seconds remaining before the status must be changed to LOST if no further HELLO message is received
Sequence	The value of the sequence number in the last HELLO message received on interface I from interface J
Count	The remaining number of times a Neighbor Request, Reply, or Lost message containing J must be transmitted on interface I
History	A list of the sequence numbers of the last HELLO_ACQUIRE_WINDOW HELLO message received on interface I from interface j
Metric	An optional measure of the quality of the link I,J represented by an integer between 1 and 255, in which smaller values indicate better quality
Priority	The relay priority of the node associated with interface J

Neighbor Table

As previously discussed, each node maintains a neighbor table for each of its local interfaces. Entries in the neighbor table are obtained from HELLO messages received by an interface. Table 4.1 lists the possible entries in the neighbor table for interface I for its neighbor J.

In examining the entries in Table 4.1 note that the value of the HELLO sequence number is used to determine the number of HELLO messages that have been missed. Also note that an entry for interface J in the neighbor table for interface I may be deleted if no HELLO was received on interface I from interface J within twice the last hold time. The hold time, which is expressed in seconds, is computed while NEIGHBOR LOST messages containing J are transmitted. In addition, the absence of an entry for a given interface J is equivalent to an entry with a status value of LOST and a history value of NULL.

Status Values

As indicated in Table 4.1, the status value for interface I to J can have one of three values: LOST, 1-WAY, or 2-WAY. A value of LOST indicates that interface I has not received a sufficient number of HELLO messages from interface J. In comparison, 1-WAY indicates that interface I received a sufficient number of HELLO messages from interface J but the link is not 2-WAY. The third possible value, 2-WAY, indicates that interfaces I and J both received a sufficient number of HELLO messages from each other.

Transmitting HELLO Messages

Under the TBRPF protocol each node must transmit on each of its local interfaces at least one HELLO message per HELLO_INTERVAL, whose default value is one second. HELLO messages can be transmitted at more frequent intervals, however, the time between two consecutive HELLOs on a given interface must be greater than the NBR_HOLD_TIME parameter divided by 128. This avoids the possibility that the HELLO sequence number wraps around to the same value prior to a neighbor that stops receiving HELLO messages changes the status of the link to LOST. The default value of the NBR_HOLD_TIME parameter is three seconds.

Because the synchronization of control messages can result in collisions HELLO messages should not be transmitted at equal intervals. Instead, nodes select an interval randomly with a value up to the HELLO_INTERVAL value. Each HELLO message always includes a NEIGHBOR REQUEST message, where the latter includes the sequence number that is incremented by 1 (module 256) each time a HELLO is sent.

The HELLO message will also include a NEIGHBOR REPLY message if its list of neighbor addresses is not empty, whereas a NEIGHBOR LOST message is included if its list of neighbor addresses is nonempty. The actual contents of the NEIGHBOR REQUEST, NEIGHBOR REPLY, and NEIGHBOR LOST messages depend upon the setting of the status and count entries in the neighbor table. Table 4.2 indicates how the contents of the three messages are determined based upon the values of the previously mentioned neighbor table entries.

Table 4.2   HELLO Message Content Determination

Neighbor Status	Table Entries Count	HELLO Message Content
LOST	>0	Include J in NEIGHBOR LOST message and decrement count
1-WAY	>0	Include J in NEIGHBOR REQUEST message and decrement count
2-WAY	>0	Include J in NEIGHBOR REPLY message and decrement count

An exception to the operations listed in Table 4.2 occurs when a node restarts. When this condition occurs all entries are purged from the neighbor table. In addition, the node ensures that for each interface at least one of the following two conditions is met:

- The difference between the transmission times of the first HELLO sent after restarting and the last HELLO before restarting is at least twice the Number_Hold_Time in seconds.
- Let the parameter value of HSEQ_LAST denote the sequence number of the last HELLO message transmitted prior to restarting and the sequence number of the first HELLO message transmitted after restarting be set to the HSEQ_LAST+NBR-HOLD_COUNT + 1 (module 256).

The purpose of either of the above two conditions is to ensure that when a node restarts each neighbor that has a link to its interfaces will set the status of the link to LOST.

Processing HELLO Messages

When a node receives a HELLO message it performs a series of predefined functions. First, it obtains the IP address of the originating interface from the IP header prefixed to the message. Next, it looks for the router ID field in the TBRPF packet header of the received HELLO message. If the HELLO message contains a router ID field, the node uses that value; otherwise, the node assigns the router ID equal to the IP address it previously obtained.

In addition to the previously mentioned operations a node performs other steps depending upon the current status of the link, the router ID value in the received HELLO message, and the presence of an entry for the interface. Such operations are described in detail in RFC 3684 which was issued in February 2004 for the experimental version of the TBRPF protocol. In addition, the referenced RFC lists the parameters used by the neighbor discovery protocol portion of the TBRPF protocol to include their proposed default values, the data structure of the topology tables maintained by each node, and the format of the topology update message. Because the RFC is experimental and the operation of the TBRPF protocol has a good probability that it will change, we conclude this section without going into the explicit details of the operation of the protocol. Instead, we conclude this chapter with an overview comparison of the AODV and TBRPF protocols.

Table 4.3 Comparing Protocol Features

Feature	AODV	TBRPF
Maximum number of nodes	1000±	200 ±
Multiple routes	No	Possible
Unidirectional link support	Possible	No
Multicast support	Possible	No

Protocol Comparison

There are several features most network managers and LAN administrators need to consider when comparing AODV with TBRPF. Those features include the maximum number of nodes each protocol can support, the ability of the protocol to support multiple routes, and unidirectional and multicast support. In addition, the ability of each protocol to be used in low- and high-mobility scenarios is important and must be considered.

Table 4.3 provides a general comparison of the previously mentioned AODV and DSR features. In addition to the four features listed in the table it is important to note that AODV is a reactive protocol, whereas TBRPF represents a proactive routing protocol.

Of the four features listed in Table 4.3 this author would consider the maximum number of nodes supported to be the most critical if you are in doubt about the potential number of clients your wireless mesh network will support. For example, when setting up a community network an initial subscriber base could rapidly expand as additional people become aware of the service. If you use TBRPF-compliant equipment it's possible that you may need to set up additional networks to support an expanding subscriber base, whereas the use of AODV-compliant equipment could support an expanded base without the necessity for establishing separate networks.

Traffic Support

When considering the ability of a routing protocol to support traffic several metrics need to be considered. Those metrics include the ratio between the amount of incoming and actually received data packets (packet delivery ratio), the end-to-end delay (latency) of packets, and the total number of control packets to data packets (routing overhead). Table 4.4 provides a general comparison between the AODV and TBRPF for low- and high-mobility scenarios, with a low traffic rate occurring under low mobility

Table 4.4   AODV versus TBRPF

Mobility/Traffic Level	Feature	Routing Protocol	
		AODV	TRBPF
Low/low	Packet delivery ratio	High	High
	Latency	Low	Medium
	Routing overhead	Low	Medium
High/high	Packet delivery ratio	High	High
	Latency	Medium	Medium
	Routing overhead	High	Medium

whereas a high traffic rate is presumed to occur under the high-mobility scenario. Note that the routing overhead is highly dependent upon the mobility mode of operation. Because AODV transmits many small routing control packets it is well suited for both low- and high-mobility operations. Similarly, because TBRPF is proactive and has a constant routing overhead it is also very suitable for both low- and high-mobility operations.

Latency of AODV for low mobility and low traffic is slightly better than TBRPF, because AODV uses small routing control packets. However, for high-mobility and high-traffic operations, latency associated with either protocol becomes very similar. The last comparison shown in Table 4.4 concerns routing overhead. Although AODV uses smaller routing control packets than TBRPF, the former is only more efficient when supporting low-traffic environments. As traffic increases TBRPF, which represents a proactive protocol, becomes more efficient as it has a constant routing overhead.

Chapter 5

Network Operation

Although a discussion of the characteristics of wireless mesh networks and an understanding of routing protocols are important, it's of equal importance to understand their potential utilization. In this chapter we examine how several wireless mesh networks can be used to support a variety of networking activities. However, because of the lack of mesh networking standards, we examine network operational examples associated with specific vendor products instead of by application. The rationale for this method of presentation of network operational examples results from present interoperability problems among different vendor equipment. It is hoped that within a few years standards will be promulgated which, when followed, will enable different vendor products to interoperate.

A second reason for focusing upon wireless mesh networking operational examples by vendor is due to the differences in hardware and software among vendor products. For example, some vendor products are designed to interconnect wired LANs or individual clients connected to routers, and other products enable individual wireless clients to participate directly in a wireless mesh network. By examining network operations with the viewpoint of a single vendor at a time, we obtain a better appreciation for the similarities and differences between currently available wireless mesh networking products. That said, let's turn our attention to wireless mesh networks created through the use of single vendor products.

5.1 Working with Firetide Equipment

Firetide is a relatively new networking company which was founded during 2003. The company currently has offices in Honolulu, Hawaii and Los Gatos, California. Firetide manufactures two types of wireless mesh routers under the HotPoint label. One router, referred to as the HotPoint 1000S is designed to be used indoors, and the vendor's HotPoint 1000R represents a wireless mesh networking router developed for outdoor applications. The key difference between the two devices in addition to the outdoor device being able to handle inclement weather concerns their Ethernet port support. The indoor HotPoint router has three Ethernet ports whereas its outdoor cousin has two.

Router Operation

Firetide HotPoint mesh routers can be viewed as an enhancement to a typical wireless access point. As a review, an access point serves wireless clients and has a connection to a wired network as illustrated in the top portion of Figure 5.1. In comparison, a Firetide HotPoint wireless mesh

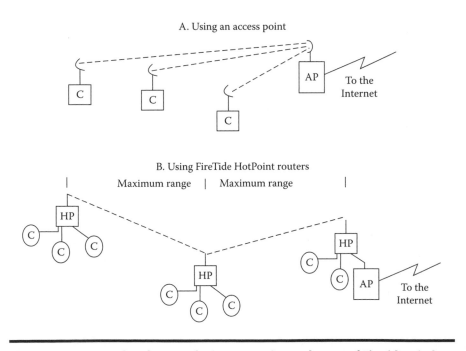

Figure 5.1   Comparing the use of an access point to the use of Firetide wireless mesh routers.

router can be viewed as providing an inverse networking function, enabling Ethernet switches or individual clients to be wired to the router which in turn provides a wireless connection to other devices in the network.

Routing Protocol Support

HotPoint wireless mesh routers use IEEE 802.11 protocols for the transmission of data. However, for routing they use a Firetide enhanced version of the Topology Broadcast based on Reverse-Path Forwarding (TBRPF) protocol to manage the mesh network formed by multiple routers. An overview of this routing protocol was presented in the second section of Chapter 4. A portion of the enhancement to the TBRPF routing protocol involves a patented technology that enables HotPoint routers to automatically locate one another and form a mesh networking structure.

Hardware Features

Each HotPoint wireless mesh router includes either two or three 10/100 auto-sensing Ethernet ports. Each Ethernet port can support the connection of an individual wired client, the output of an Ethernet switch, or an AP connected to a wired network. As previously noted, HotPoint mesh routers use a proprietary version of the Topdogy Broadcast based on the TBRPF routing protocol to support their mesh networking capability.

Network Utilization

The lower portion of Figure 5.1 illustrates the use of three HotPoint wireless mesh routers to create a mesh network that enables one to three individual clients to be networked and obtain access to a wired network's Internet connection via an access point connected to a port on one HotPoint router.

In examining the top portion of Figure 5.1, note that the client shown on the left of the illustration is outside the maximum communications range of the access point. Due to this, that client cannot establish communications with the access point. In comparison, the two clients within range of the access point can communicate with that device and use its bridging capability to obtain a connection to the Internet.

If we turn our attention to the lower portion of Figure 5.1 we can obtain an appreciation for some of the advantages associated with the use of wireless mesh networking devices by examining how the three Firetide HotPoint wireless mesh networking routers are used. Because the maximum transmission range would not enable the leftmost HotPoint router shown in Figure 5.1 to be able to communicate directly with the access point, that router uses the middle HotPoint wireless router as an intermediate device. Thus, communications from the leftmost router are relayed via the middle HotPoint router cabled to the access point. This illustrates how the use of wireless mesh networking equipment can expand the range of coverage of a wireless LAN. As additional HotPoint routers are added to the network, the use of their locations as intermediate devices can result in an additional transmission range becoming available. This occurs because clients connected to the routers become able to access the Internet through the extension to the range of the mesh network.

Concerning those clients, at the time this book was written Firetide HotPoint routers were limited to supporting either two or three conventional 10/100 Mbps Ethernet ports, with the number of ports varying by the type of router. As previously mentioned, the outdoor router supports two ports and the indoor router supports three. Because many conventional wireless access points include Ethernet switch ports, this author called the vendor to determine if Firetide had any plans to upgrade their HotPoint wireless mesh routers to include a built-in port switching capability that would enable a network segment to be connected to a switch port. Although the vendor mentioned that they were considering the addition of this capability, they did not have any anticipated date for which the addition could be expected.

Security

One of the more interesting aspects of the HotPoint wireless mesh routers concerns its security for over-the-air transmission. HotPoint routers can be configured to communicate with one another using triple DES encryption. For additional protection, all HotPoint routers use unique digital certificates as a mechanism to authenticate each other on the network. Through the use of authentication and triple-DES encryption, HotPoint routers remove the security vulnerabilities associated with the use of a network name, better known as the Service Set Identification (SSID) and Wired Equivalent Privacy (WEP) used with conventional wireless LANs. In fact, at the time this book was written HotPoint had added support for the Advanced Encryption Standard (AES) and allowed the layer 2 link to be configured to use WEP even though AES was sufficient to secure transmission. Thus, although the Firetide HotPoint wireless mesh routers use IEEE 802.11-compliant transmission to form a wireless RF-based network, their security is hardened in comparison to the first generation of wireless LAN products that reached the market.

Modes of Operation

Firetide HotPoint wireless mesh routers can operate in one of two modes which define the type of network that can be constructed. The two modes are public and private and the selection of a specific mode of operation governs the capability of the device to perform routing and forwarding. When placed in a public mode of operation Firetide routers forward any traffic to and from destinations within or beyond the network. Thus, you would configure HotPoint routers to operate in their public mode if you wanted clients to have the ability to reach the Internet or a wired network connected to an access point which in turn was cabled to a port on a Firetide router that was used to form the mesh network. In comparison, if the routers are configured for a private network mode of operation, packets will only be forwarded to or from destinations within the private network. Thus, setting the Firetide HotPoint routers to a private network mode of operation precludes the private network from being used as a transport facility for public networks. However, such transport restrictions are only applicable to the routing of messages and are not necessarily a characteristic of client nodes connected to the HotPoint routers.

To illustrate the differences between HotPoint public and private network operations consider Figure 5.2. That illustration shows four wireless networks, with two set to a public network mode of operation and the other two set to a private network mode of operation. In this example traffic from node A cannot traverse either private network to reach node D.

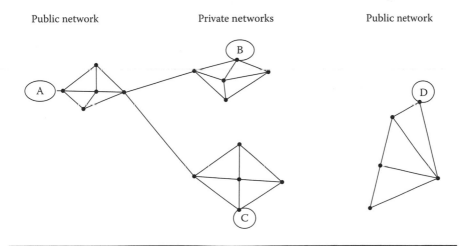

Public network   Private networks   Public network

Figure 5.2   By placing routers in a public or private mode of operation, you can control the flow of packets in a HotPoint mesh-formed network.

However, traffic from node A can communicate with nodes B and C because the latter two nodes are endpoints within their private networks. Thus, if node D were an access point connected to the Internet, traffic from node A on the public network would not be capable of reaching the mother of all networks. However, changing either of the two private networks shown in Figure 5.2 into a public network would then allow node A to communicate with node D.

Outdoor Utilization

The Firetide HotPoint 1000R is a wireless mesh router designed for outdoor operations. This router includes two weatherproof 10/100 Ethernet connectors and a remote power module with a 100-foot power cable. An external 24-inch rod attaches to the unit via a connector and cable at the top of the router's enclosure. The entire enclosure can be mounted on a wall or pole.

The HotPoint 1000R outdoor mesh router uses a 200-mW radio. This enables line-of-sight transmission distances of up to two miles to be achieved, making the use of outdoor routers applicable for interconnecting buildings within an office park or campus. For example, consider Figure 5.3 which illustrates a five-building office park. In this example each building has a Firetide HotPoint 1000R outdoor router mounted on a roof that provides a line-of-sight transmission capability to other antennae mounted on the roofs of other buildings in the office park. If we assume a LAN switch is cabled to a port on each router it becomes possible to interconnect numerous clients within each building into a wireless mesh network that spans all five buildings in the office park.

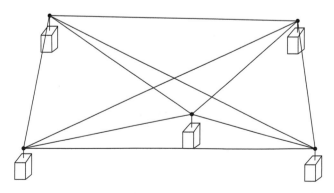

Figure 5.3 Using outdoor Firetide HotPoint routers to create a mesh network within an office park.

5.2 LocustWorld Mesh AP and Meshbox Networking

LocustWorld represents a United Kingdom-based organization that was one of the pioneers in understanding the potential of free community networks. Currently LocustWorld offers hardware in the form of its MeshBox and MeshBox 2X as well as client software that can be used to create community networks. Both hardware and software use the AODV protocol to create a mesh networking environment. Each mesh network is autonomous, discovering routes on demand with unused routes eliminated from routing tables after a predefined period of time. In the following sections we examine LocustWorld's client software, which is referred to as MeshAP, as well as the organization's two hardware products, which are referred to as the MeshBox and MeshBox X2.

MeshAP

At the time this book was written MeshAP software was available for free from http://locustworld.com. This software was developed to enable IEEE 802.11 wireless LAN-enabled hosts to scan the airwaves and join other nodes in a mesh-enabled network, in effect extending the network. The software is fully automatic, requiring no setup process nor configuration of hardware or software. The entire network can be peer-to-peer, because no central infrastructure in the form of an access point is required. To obtain an appreciation for the manner by which MeshAP software provides automatic, configuration-free operation let's probe further and describe how its software design facilitates this method of operation.

Software Design

Mesh AP software operates on a wireless LAN-enabled host. That host then becomes a node in a mesh-enabled network. If the node has a connection to the Internet, it will function as a gateway node; otherwise, it functions as a repeater or an extension to the mesh network.

As the software known as MeshAP boots, it allocates itself an RFC 1918 IP address, typically in the 10.X.X.X range. Initially the IP address is allocated randomly. Once an IP address is allocated the MeshAP software attempts to locate an internal gateway. In doing so it probes the RFC 1918 address block 192.168.1.1/255.255.128.0 and then using a DHCP client on the Ethernet interface of a located gateway attempts to log into the device. If the software cannot locate a gateway via its probe of the previously mentioned RFC 1918 address block, it considers the host it is residing on as the only wireless link and thus becomes a repeater cell. The repeater cell then starts an internal DNS server and transparent Web proxy on port 80 and port 8080, a DHCP server is started, and a random RFC class C network is selected in the range 192.168.128.0/255.255.128.0.

As other Mesh AP clients become active and connect to the DHCP cell they will be pointed to the wireless interface for the default gateway and DNS server. In fact, the DNS server operating on the repeater is designed to always return the address of the gateway regardless of what domain name is resolved. At this point the node functioning as a repeater cell can sign clients on but cannot serve such clients inasmuch as a routing protocol is required.

As the repeater cell node signs clients on, the AODV module is loaded, enabling the node to find neighbor cells within its wireless LAN transmission range by transmitting and receiving User Datagram Protocol (UDP) packets to the network broadcast address. Because any cells within the mesh that are gateways periodically broadcast a route to a bogus address which is used to define an Internet gateway, monitoring for that address enables gateways in the mesh network to be located.

When a cell without a gateway receives the bogus address it will attempt to establish an encrypted IP tunnel using compression and blow-fish encryption. The IP tunnel can occur over multiple hops to the destination gateway, with the AODV protocol being responsible for the routing between linked cells. The cell will then switch all its outbound DNS and IP traffic to flow over its recently established VPN connection to the gateway. The DHCP configuration will also be updated to function as the remote gateway address and as a DNS server. Any clients who signed on to the node prior to the link to the gateway being created will forward traffic to the local cell. That cell will function as a proxy, transmitting information via HTTP. In comparison, any clients signing on after the gateway is located will receive services from the remote gateway and have full IP routing capability. In addition, any client signing on to a cell that has a local Internet gateway capability will be able to access the Internet directly via that gateway.

Limitations

Similar to many software programs, the Mesh AP has a number of limitations. Some of the current limitations concern the operation of the software and other limitations involve the omission of security features that would be more than likely required prior to the use of the program for economic-related activities.

Software Operation

One of the key limitations of the Mesh AP program is only noticeable in large networks. This limitation is referred to as the absence of a deadlock avoidance mechanism associated with IP address allocation. Because IP addresses are selected randomly, as the number of nodes in a mesh network increases so does the probability that two or more nodes could randomly be assigned the same IP address, a situation that the developer is aware of and which can be expected to become the subject of additional effort.

A second operational area that needs further effort concerns the use of multiple gateways. In an ideal environment the mesh network should have a mechanism to detect loading over the links and allow for the optimization of routing based on current loading and probability to enable full redundancy to be achieved.

Security

The current version of Mesh AP software has several security related limitations. First is the absence of authentication, which allows a total stranger to download Mesh AP software and become a participant on a Mesh AP-created mesh network.

A second security limitation of the Mesh AP software is the absence of cryptographic registration of nodes. Because participants in a mesh network tend to commence operations with a predefined key, unless the key is changed, users downloading Mesh AP software could use default settings to establish a secure tunnel to a gateway.

In addition to the previously mentioned security issues, the use of the AODV routing protocol has a limitation—the absence of a checking mechanism when table updates are transmitted—which means that it's possible for a third party to generate false entries which, when transmitted to neighbors, have the effect of causing potential routing havoc within the network.

Software Availability

If you want to create a Mesh AP multi-hop wireless mesh network you can obtain open source software from http://www.locustworld.com and via the academic mirror at ftp://ftp.mirror.ac.uk/sites.locustworld.com/. From either location you can download a Zip file of approximately 28 Mbytes that includes both a "README" file and a file that, when written onto a CD, becomes a bootable CD that will work in many laptop and desktop computers.

At the time this book was written Version 06 of the Mesh AP was the latest version available for download. This version supports several IEEE 802.11 wireless LAN adapter cards to include some of those manufactured by Lucent, Orinoco, and Buffalo. In addition to creating a mesh network you can also use this software to turn your laptop or desktop into a wireless access point or even use your computer to test if the computer is within range of an existing Mesh AP-based wireless mesh network.

MeshBox

The LocustWorld MeshBox represents a self-contained IEEE 802.11b wireless LAN communications-based computer system that is approximately the size of a small video recorder. The MeshBox includes a single board computer and 128 Mbytes of memory as well as 32 Mbytes of nonvotile storage, dual USB ports, an Ethernet port, serial and parallel ports, and an antenna.

The MeshBox computer operates a custom version of the Linux kernel, with mesh routing and interactive functions provided as Linux applications. Through the use of the AODV routing protocol, the MeshBox communicates with other MeshBoxes within range of each other to provide a self-organizing and self-healing network with dynamic routing between nodes. Clients operating IEEE 802.11b or 802.11g adapters can connect to one MeshBox and then obtain network services from other mesh nodes, such as a broadband Internet connection.

Each MeshBox can be connected to a broadband Internet service either via the computer's RJ45 Ethernet port or one of its two USB ports. Connectivity can occur via a cable modem, DSL connection, or even a wired Ethernet connection. Through the use of an omnidirectional antenna and built-in IEEE 802.11b communications card, each MeshBox provides an indoor transmission range of approximately 100 meters (300 feet) and an outdoor range of approximately 400 meters (1300 feet).

Applications

One of the key application areas for the MeshBox is to provide affordable broadband Internet access for individuals located outside metropolitan areas where cable TV and DSL service may not be available or only available at relatively high prices. Through the use of LocustWorld MeshBoxes, Wireless Internet Service Providers (WISPs) can provide public access wireless broadband services.

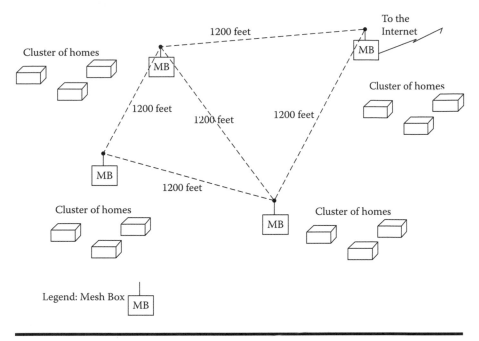

Figure 5.4   Using LocustWorld MeshBoxes to provide rural area Internet connectivity.

WISP

Figure 5.4 illustrates one example of the use of LocustWorld MeshBoxes by a wireless Internet service provider to extend Internet connectivity to a rural area. In this example, the WISP placed three MeshBoxes within 1200 feet of one another to serve three clusters of homes within a rural area, with a fourth MeshBox used to extend transmission to an area where the service provider has a broadband connection to the Internet. In this example clients in each home connect to a MeshBox and are routed to the Internet, with only one high-speed broadband connection required to support many users. For the example illustrated in Figure 5.1, MeshBox 1 must have its data relayed via MeshBox 2 or MeshBox 3 to reach the Internet. In comparison, MeshBox 2 and MeshBox 3 clients can directly access MeshBox 4 and its high-speed Internet connection or MeshBox 2 can use MeshBox 3's connection to MeshBox 4, and MeshBox 3 can use MeshBox 2's connection to MeshBox 4 and gain Internet access.

In addition to being used by WISPs, MeshBoxes can be set up in a business environment, within an office park, or on a college campus. However, despite the diversity of potential applications, the MeshBox has received a preponderance of publicity for its use in creating community networks.

Community Networking

One example of the use of MeshBoxes for the establishment of a community network recently occurred in Devon, England. Devon is located in a part of South Hams, Kingsbridge that is similar to many other areas in rural portions of the United Kingdom with respect to the availability of broadband Internet access. Having been passed over by the DSL service provider, the Devon community pooled their resources to acquire MeshBoxes from LocustWorld to create an adaptive, ad hoc, and self-healing mesh network with a single broadband connection instead of having computer users in the village individually investing in separate Internet connections. The apparent success of Devon resulted in other communities in the area, such as Gartocharn and the village of Langtoft in East Yorkshire either considering establishing wireless mesh networks or using MeshBoxes to establish a network. Concerning the latter, the village of Langtoft used LocustWorld MeshBoxes to provide wireless broadband access to over 400 residents of the village and surrounding community.

Through the use of a 1-Mbps Aramiska satellite Internet connection and 8-dB omnidirectional antennae on MeshBoxes a wide area of coverage was obtainable. The selection of 8-dB omnidirectional antennae instead of high-gain devices enabled coverage from roofs to customers on ground floors to be provided. Otherwise, the use of high-gain antennae would restrict coverage to users on upper floors of buildings where antennae were mounted on roofs. Other communities, such as Southhampton, Edinburgh, Cardiff, and Kent were experimenting with wireless LANs at the time this book was prepared. Since its launch in November 2002, LocustWorld has sold in excess of 200 MeshBoxes in the United Kingdom to communities who want to share a broadband connection over a large number of customers, in effect significantly reducing the cost of broadband Internet service from the individual broadband service model that telephone companies and cable systems prefer to offer customers.

Contract Constraints

Although end users in communities offering MeshBox service are happy with the economic savings, communications carriers are not exactly thrilled. Because U.K. communications carriers are similar to their U.S. cousins in the fact that the sharing of a residential broadband connection outside the address being served is against the terms and conditions of their contracts, the use of MeshBoxes or similar products could wind up in court. However, because communication carriers also market T1 and European E1 Internet connections for commercial use, it's quite possible that communities could order a T1 or E1 service terminated at a public building and use MeshBoxes or similar devices to establish a wireless mesh network over a large geographical area without violating the terms of a communications carrier's contract.

Security

The use of MeshBoxes to form a wireless mesh network includes several operational features beyond standard WEP to provide security. First, MeshBoxes support 2048-bit certified encryption throughout the formed mesh network. In doing so all traffic between MeshBoxes is encrypted via Point-to-Point Tunneling Protocol (PPTP) connections. In addition, last-hop traffic can be encrypted with a VPN connection to the local node because each MeshBox supports passthru for both the use of PPTP and IPSec.

In concluding our discussion of the LocustWorld MeshBox this author would be remiss if he did not mention the MeshBox X2. The MeshBox X2 represents an updated MeshBox that includes two IEEE 802.11b adapter radio modules. At the time this book was written it appeared that the new MeshBox would be marketed under the name Mexbox although the term MeshBox 2X was still being used on the vendor's Web site to refer to this product.

The goal of the MeshBox X2 or Mexbox is to support large mesh networks in terms of client node connections. By supporting dual wireless LAN adapter cards, the MeshBox X2 provides, in effect, a doubling of bandwidth in comparison to the use of the original MeshBox. Because there are 11 channels available for use in North America and 13 in Europe, upon setup each box selects two of the allowable frequencies while making sure that at least one is the same as any nearby MeshBox or MeshBox X2. If one box has channels 1 + 2 and a second MeshBox X2 has channels 2 + 3 in operation while a third box has channels 3 + 1 in operation, then routing can go from 1 to 2 to 3 and back to 1. Through the use of the Meshbox X2, the capacity of a wireless mesh network is significantly expanded, enabling more users to simultaneously use the network to obtain a broadband Internet connection. In fact, a customer of LocustWorld located in the Caribbean was planning to construct a rural broadband network in which the use of the MeshBox X2 would enable the community network to operate with a reduced number of MeshBoxes due to the added capacity of the newer device.

Economics

At the time this book was written the cost of a MeshBox was approximately $250, with the price of Meshbox X2 double that. According to LocustWorld, a commercial mesh network becomes viable with approximately 50 domestic or 20 business users. Because a co-op mesh is operated by volunteers which results in a lower operating overhead, according to the company such co-op domestic use wireless mesh networks can become economical with approximately 20 users.

To determine the economics associated with establishing a wireless mesh network for a community, you need to consider the cost of three key elements: the broadband Internet connection, the number of MeshBoxes required to service an area and their cost, and the cost of personnel to provide billing and customer support. To illustrate the economics associated with establishing a community network let's assume two Internet connection scenarios. In the first scenario we assume the use of a cable modem or DSL connection is available for $40 per month and the communications vendor will allow the connection to be shared by multiple households. In the second scenario we assume that due to contract restrictions the community network operator must install a T1 or E1 circuit to a community building at a cost of $1000 per month. It should be noted that in some locations it may be possible to have the local telephone company install a symmetrical HDSL circuit to provide the Internet connection. If an HDSL circuit is available, the monthly circuit cost could drop to approximately $500.

Continuing our economic model analysis, let's assume we need to cover a geographic area of approximately one-half square mile. Because a MeshBox can provide an outdoor transmission distance of approximately 1200 feet, the use of nine MeshBoxes would be sufficient to ensure that all clients within the half-square mile location could access the mesh network through at least two boxes. Figure 5.5 illustrates how the positioning of nine MeshBoxes could provide a redundancy of coverage that extends over a geographic area of approximately one-half square mile, an area that could easily provide service to several suburban neighborhoods.

Because the cost of a MeshBox including its antenna is $500 per unit, for either scenario the total cost becomes $500 × 9 or $4500. If we assume that community network organizers will provide their time for billing and customer support as volunteers, then there is no cost associated with their effort. Table 5.1 compares the cost of the two scenarios for the first two years of operation.

In examining the upper portion of Table 5.1 comparing the cost of community networking for cabled DSL service and T1/E1 Internet access

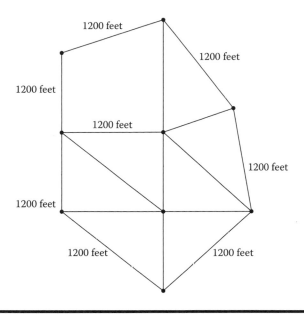

Figure 5.5 **Using nine MeshBoxes provides redundancy for clients over a half-mile geographic area of coverage.**

Table 5.1 **Community Networking Cost for Coverage of a Half Square Mile**

	Cable/DSL Service		T1/E1 Required	
	First Year	Second Year	First Year	Second Year
Internet access	480	480	12000	12000
MeshBoxes	4500	0	4500	0
Support	0	0	0	0
Total cost	$4980	480	16500	12000
Cost/Month Projections				
25 users	16.60	1.60	55.00	40.00
50 users	8.30	.80	27.50	20.00
75 users	5.53	.53	18.33	13.33
100 users	4.15	.40	13.75	10.00

you will note a considerable difference between the two. Unfortunately, due to most Internet service provider contract terms, you will normally be precluded from sharing the use of a cable or DSL Internet connection among multiple homes. Thus, a more realistic cost for establishing a community network would be associated with the use of a T1 or E1 access line.

Even though the cost associated with the use of a T1 or E1 access line is considerably more than that required when a cable or DSL service connection is used, that cost is only part of the story. To understand the rest of the story we need to compute the yearly total costs based upon the number of users within the community that could take advantage of the community network. Thus, the lower portion of Table 5.1 provides a more relevant economic projection because it amortizes the total yearly cost estimate by different numbers of potential community users. In the lower portion of Table 5.1 this author divided the total cost projections for each scenario's first and second year of operation by four different user groups to obtain four monthly cost estimates. As indicated in the lower portion of Table 5.1, even when a T1 or E1 connection is required, a community network base of 100 users can reduce the cost of broadband Internet access to less than the fees modem users spend on dial-up 56-kbps service.

In addition, it's important to consider the fact that telecom rates are in almost a free-fall rate of descent. According to published reports in trade journals during 2005, users can expect to get between 25 and 30 percent price reductions when negotiating a new contract for the same service. In fact, some communications carriers were offering a 200-mile T1 Internet connection for under $500 per month if the purchaser was willing to sign a three-year agreement. If you revise the costs shown in Table 5.1 by halving the yearly cost of a T1 circuit then the cost projection to service 100 users would fall well below $10.00 per month per user during the first year of community networking and to $5.00 per month per user for the second and subsequent years of operation! As the commentator Paul Harvey would say, "Now you know the rest of the story."

5.3 Packet Hop Networking

A spinoff from SRI International, Packet Hop is a Belmont, California-headquartered startup that was formed during 2003. In February of 2004 Packet Hop delivered a mobile mesh solution for the homeland security market that enabled communications connectivity via mesh networking on and in the vicinity of the Golden Gate Bridge to include support for locations on the water as well as on land around the bridge.

Network Components

The Packet Hop mobile mesh network is formed through the use of software that is designed to operate on commercial off-the-shelf hardware. There are three components used to form a Packet Hop mesh network: a network controller, a network management system, and network clients.

Network Controller

The Packet Hop network controller can be considered to represent a sophisticated access point that interfaces the mobile mesh network to an organization's existing wired network. Besides acting as a bridge between the wired and wireless world, the network controller manages connectivity between multiple mobile mesh networks, provides network managers with the ability to set policies via the use of access control settings, and enables roaming, security, and Quality of Service (QoS) among the various IP networks being controlled.

Network Management System

The second major component of the Packet Hop wireless mesh network is the Network Management System (NMS). The purpose of the Packet Hop NMS is to enable the mesh network to be managed and provisioned from either a central location or via multiple locations within the network. The NMS provides managers with real-time information concerning the operational status of the mesh network as well as the ability to set network policies and troubleshoot problems.

Network Clients

The third component of a Packet Hop mesh network is its network clients. Packet Hop software operates in a Windows environment and supports IEEE 802.11 wireless LAN products. This means that a Packet Hop wireless mesh network can be established through the use of IEEE 802.11-compatible RF equipment installed on laptops, PDAs, tablet computers, and even wearable computers as long as such computers are Windows-compatible.

Multi-Terrain Support

In February, 2004 Packet Hop equipment was used by the Golden Gate Safety Network in an exercise to establish mobile broadband connectivity both on and around the Golden Gate Bridge in the San Francisco Bay area. In a live field exercise Packet Hop equipment was used to provide communications that enabled over ten multijurisdictional city, state, and federal agencies to be interconnected. Because some participating agencies, such as the U.S. Coast Guard used Packet Hop equipment in the San Francisco Bay area and other agencies used equipment on the Golden Gate Bridge and surrounding land areas, this exercise also demonstrated the feasibility of creating a mixed-terrain wireless mesh network.

Network Applications

Because Packet Hop network components support QoS among IP networks the network can provide support for many types of applications. Such applications include the transmission of real-time multicast video and multimedia messaging. Multicast video can include fixed-to-mobile as well as mobile-to-mobile video multicasting. Thus, video taken from a Coast Guard ship located off the Golden Gate Bridge could be transmitted to clients on land and members of agencies patrolling the bridge on land could transmit video of suspicious activity to other agencies within the vicinity of the Golden Gate Bridge that were participating in the wireless mesh network. Concerning multimedia messaging, Packet Hop supports instant messaging with images and video for peer-to-peer transmission as well as for provisioned user groups. This means that in the previously mentioned Golden Gate Bridge exercise, different city, state, and federal agencies that participated in the wireless mesh network were capable of being formed into different user groups to transmit instant messages among group members including sending images and video when required.

5.4 Nortel Networks Wireless Mesh Networking

Nortel Networks appeared to be the only major communications company to offer wireless mesh networking products when this book was written. Marketed under the "wireless mesh network solution" banner, Nortel offered a series of three interrelated products developed to extend IEEE 802.11 wireless LANs into a mesh networking environment. In addition, although not a member of the troika of solution products, Nortel also markets an optical wireless network solution based upon the mesh networking of free space optical links which deserves mention and is covered in the second half of this section. However, prior to examining the Nortel optical system, let's focus our attention upon the company's wireless mesh networking products that use IEEE 802.11b and g interfaces.

IEEE 802.11b/g Products

Nortel Networks' wireless mesh network solution is formed through the use of three main network components. Those components include the firm's 7220 wireless access point, 7250 wireless gateway, and its Optivity network management system.

7220 Access Point

The Nortel Networks 7200 wireless access point is a most interesting product as it uses two Industrial, Scientific, and Medical (ISM) frequency bands. The access point uses the 2.4-GHz frequency band to support transmission to mobile nodes in what is referred to as the access point's access link (AL). Between access points, communications occur in the 5-GHz ISM band over what is referred to as the access point's Transmit Link (TL).

The 7220 access point is designed to be hung from an existing utility pole or wall mounted so that the AP is elevated. The access point uses a dual-polar antenna with transmission employing IEEE 802.11a technology on its TL link. Because the wireless AP performs traffic collection and distribution functions it incorporates a routing capability. In addition, according to Nortel literature, the 7220 access point also includes security functions that are used for validating connections to other wireless APs and controlling access to APs by mobile nodes.

7250 Wireless Gateway

A second major component of a Nortel Networks wireless mesh network is the firm's 7250 wireless gateway. A key function of the wireless gateway is to advertise reachability within an enterprise or an ISP for different subnets assigned to community area network subscribers and network entities. Because data flow from the Internet and wired LANs first flows to the 7250 wireless gateway, that device in effect hides community area network-specific mobility from the backbone wired infrastructure. Figure 5.6 illustrates a possible hierarchy of Nortel Networks wired and wireless equipment that can be used to span a geographic area comprised of several community area networks that are linked by an ISP to the Internet.

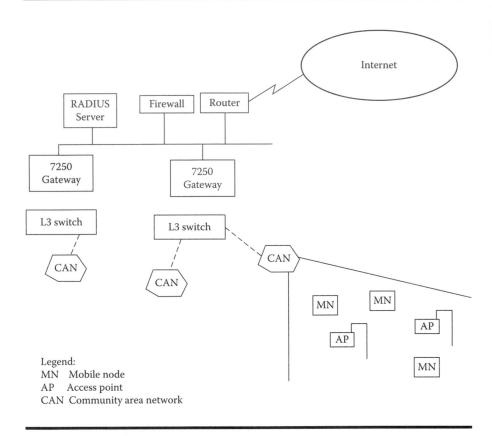

Figure 5.6   Using a hierarchy of Nortel Networks equipment to interconnect several community area networks to the Internet.

Note the possible use of a RADIUS server and firewall on the wired portion of the network that extends connectivity to the Internet.

Optivity NMS

A third component of the Nortel Networks wireless mesh networking solution is the vendor's Optivity Network Management System (ONMS). Under version 10.2, support was added for wireless LAN devices. This support includes a discovery feature which enables the ONMS to discover wireless LAN devices and display the discovered topology. Other wireless mesh networking support built into Optivity NMS 10.2 includes organizing all mesh devices into a common folder to facilitate their organization, monitoring the performance of wireless mesh network devices, and performing fault management via the use of wireless mesh device traps. Under ONMS 10.2, support for the Nortel 7220 wireless access point and

7250 wireless packet gateway is provided in addition to the vendor's conventional 2220 and 2221 wireless LAN access points and 2250 wireless LAN security switch. Thus, the use of the ONMS permits both conventional wireless LANs and wireless mesh networks to be managed. In addition, because the ONMS also supports the management of wired devices it can be used on an enterprise basis to manage both the wired and wireless infrastructure. Now that we have an appreciation for the major Nortel Networks wireless mesh networking components we conclude our discussion of this vendor's products by turning our attention to its open-air system designed for "last mile" access.

The OPTera Metro 2400

One of the major problems associated with "last mile" provisioning is the cost associated with establishing connectivity to new buildings in suburban areas. When an office or apartment complex is built in a rural area it can be quite expensive for a communications carrier to extend service to the new occupants. This is especially true when the existing central office or cable network branch is located beyond the support range of new buildings, requiring the communications carrier to invest a substantial amount in creating an extended infrastructure. Recognizing this problem, Nortel Networks introduced its OPTera Metro 2400 open-air system. This optical wireless device supports mesh networking of free space optical communications links at a full duplex data rate of 622 Mbps with a planned upgrade to a data rate of 2.4 Gbps.

The OPTera Metro 2400 is designed to be mounted on the top of buildings in a suburban or office park environment or even an apartment house where a high data rate is needed to support the communications requirements of the occupants of the building. Each OPTera Metro 2400 functions as an access point node for the building it is mounted on as well as a relay point for traffic originating elsewhere in the mesh. Because transmission is optical there is no need for a spectrum license and the high data rate enables the capacity of optical fiber to reach buildings where the extension of fiber could be costly and time consuming.

The transmission range of the OPTera Metro 2400 is up to approximately 500 meters. Because it's possible that a series of OPTera Metro 2400s mounted on rooftops might not extend to a carrier's nearest facility, the pole mounting of additional line-of-sight devices can be used to extend the range of the formed optical mesh network. To support the management of a group of OPTera Metro 2400 open-air systems Nortel Networks offers an Element Management System. This system, as you might surmise, is used to facilitate the management of the mesh network and each of its nodes.

Chapter 6

Creating a HotPoint-Based Mesh Network

Many years ago my old engineering professor was fond of saying that "The proof of the pudding is in the eating." His use of the well-known adage was always accompanied by a short experiment that illustrated an important concept. In this book on wireless mesh networking, I would be remiss if I did not follow my engineering professor's adage and demonstrate the operation and management of a wireless mesh network. In doing so the use of mesh-enabled equipment is obviously an essential part of any operational demonstration. Thus, this author gratefully acknowledges the shipment of Firetide HotPoint wireless mesh routers whose use enabled his home to be turned into a wireless community network. Firetide provided this author with three HotPoint 1000S wireless mesh routers that were used to establish his in-home community network. Prior to describing the operation of the routers, a bit of information concerning the author's home is in order, as its physical layout and wireless transmission limitations illustrate how a mesh-enabled network can satisfy a variety of communications requirements.

6.1 The Networking Environment

As briefly mentioned above, the operation and utilization of Firetide HotPoint wireless mesh routers occurred in this author's home. That home,

Figure 6.1   The author's home served as a testbed for establishing a wireless mesh network.

which is illustrated in Figure 6.1, is two stories high, with the dimensions of the first floor approximately 65 by 40 feet.

As a contemporary home, several rooms are two stories tall, however, the use of Tennessee crab orchard stone both along the front of the home and in a two-story-high fireplace in the middle of the house considerably attenuated wireless signals both to the front of the home as well as between certain rooms in the house. To obtain an appreciation of how the Firetide HotPoint wireless mesh routers were used to overcome certain networking constraints we need to first discuss the author's home networking environment. Thus, let's turn our attention to this topic.

Home Networking Environment

This author has his high-speed Internet cable modem connection terminated on a built-in desk located in the kitchen of his home. The output of the cable modem was connected to an SMC Networks wireless Barricade g router. That router has four RJ-45 10/100-Mbps built-in Ethernet switch ports for local network connections as well as a 54-Mbps wireless interface that supports up to 253 mobile users. Because the wireless router supports the IEEE 802.11g standard, it could also operate as an 802.11b device, supporting transmission of data rates ranging from a maximum of 11 Mbps down to 1 Mbps. The router's switch ports autonegotiate the operating speed to either 10 or 100 Mbps, with the mode set to half or full duplex. The pin signals on the switch ports are MDI/MDI-X, which allows the ports to be connected to any Ethernet-compatible device through the use of a straight-thru cable.

One of the SMC Networks Barricade wireless router's switch ports was cabled to the author's desktop Gateway Profile computer which was placed on the kitchen's built-in desk, leaving three ports available for connection to other network devices. The default network settings for the wireless Barricade router include a gateway IP address of 192.168.2.1 and a subnet mask of 255.255.255.0.

Because the Barricade router includes a DHCP capability this author could either configure his Gateway Profile computer's TCP/IP protocol properties to use an IP address of 192.168.2.XXX with XXX an integer from 2 to 254 or select the option "obtain an IP address from a DHCP server." Because this author periodically used up to six laptops in his home to test various software as well as equipment, it was easier to set all computers to receive their addresses from the DHCP server built into the Barricade router.

Network Access

In this author's home he could access the Internet via the SMC broadband Barricade g wireless router using either a laptop or desktop containing wireless adapters from his office on the second floor, which was located one room over from being directly above the kitchen where the router was installed. However, if this author moved down the upstairs hallway into a bedroom located at the far end of the home, both interior walls as well as the two-story stone fireplace obstructed his ability to make a wireless connection to the router downstairs. Thus, positioning one Hot-Point wireless mesh router downstairs and two located at strategic positions upstairs would appear sufficient to convert this author's home into a community network. This action would enable this author to take a laptop and achieve connectivity to the Internet from locations that were anything but ideal for reaching the router in the kitchen. As a side benefit, it should be mentioned that moving a laptop into the upstairs bedroom enabled this author to both test the use of mesh networking equipment as well as watch television, an important consideration if you're a news junkie. Now that we have an appreciation for this author's home networking environment and the potential use of Firetide HotPoint routers, let's investigate their setup, configuration, and operation.

6.2 Creating a HotPoint Mesh Network

In this section we turn our attention to the creation of a wireless mesh network in this author's home. First, we examine how HotPoint routers are set up and configured and then discuss their operation. Using the WildPackets AiroPeek program this author captures and decodes some HotPoint router packets to provide further insight concerning their operation. In concluding this chapter we note how Internet connectivity could be easily expanded to the community of homes around the author's house through the addition of HotPoint routers designed for outdoor use as well as the strategic placement of indoor routers.

Hotpoint Router Local Connectivity

Prior to positioning the HotPoint wireless routers in his home this author downloaded and installed HotPoint Manager software from the Firetide Web site as a mechanism to examine wireless mesh networking options available for configuring the vendor's routers. Version 1.6.2 of the HotPoint Manager was downloaded and installed on one of the author's Windows XP laptop computers. Other versions of the HotPoint Manager support MAC OS X, Linux, and other Java-enabled platforms. The installation of the software used approximately 42 Mbytes of storage and included a HotPoint Manager User Guide and two warranty files, with the guide and warranty information stored as Adobe Acrobat documents.

Management of the mesh network requires a hardwired connection between the platform operating the HotPoint Manager and one HotPoint router. That hardwired connection requires a straight-thru Ethernet cable. Fortunately, if your cable is not labeled you can simply look for the illumination of a green LED on the HotPoint router's Ethernet port to ensure you used the correct cable and have connectivity. In addition, because the default IP address of the mesh network router is 192.168.224.100, the computer must be on the same IP addressed network to obtain a direct connection to the router, with the Firetide manual suggesting the setting of the computer's IP address to 192.168.224.200, with a subnet mark of 255.255.255.0.

Port Connection Considerations

Each HotPoint indoor router contains three 10/100 Ethernet ports. Although you need to configure the IP address of the computer running HotPoint Manager to a valid address on the HotPoint network, when cabling a computer's LAN port to one router port, it's important to correctly configure the TCP/IP protocol properties for other computers that will be connected to the HotPoint router. For example, one of the HotPoint wireless mesh routers in this author's network will be connected to a switch port on the previously mentioned SMC Networks Barricade wireless router. That router has an IP address of 192.168.2.1 and it includes a DHCP server capability. Thus, to access the Internet either from the computer cabled to the HotPoint router running HotPoint Manager or from other computers cabled to the other HotPoint router in the network this author was establishing, you need to configure the TCP/IP protocol to either an address on the 192.168.2.0 network or select the option "obtain an IP address from a DHCP server." Otherwise, if you fail to implement either option, your browser will simply display an error message each time you attempt to view a particular Web location because your computer will not be on the correct network. Now that we have an appreciation for addressing required when cabling computers to the Ethernet ports built into the HotPoint router let's turn our attention to the use of the HotPoint Manager program.

Using HotPoint Manager

Figure 6.2 illustrates the initial Firetide HotPoint Manager log-in menu. Note that both the username and password have default settings of "admin" and should be changed after you log into the router. Also note that the Mesh IP address of 192.168.224.100 represents the default IP address of the router. Similar to the default username and password, you can also change the Mesh IP address.

Once you log into the HotPoint router, the HotPoint Manager program will display a screen that provides information about all routers in the mesh network. Because simply observing the state of the router cabled to a computer would not be very informative nor exciting, this author unboxed two additional HotPoint wireless mesh network routers. Traveling upstairs, he strategically positioned those routers at each end of a long hallway and powered them up. Returning downstairs, the use of the HotPoint Manager would now be more productive because there were now three routers operating in the mesh network.

Inventory View

Figure 6.3 illustrates the initial HotPoint Manager program screen display after you successfully log into the connected router. This screen shows the View menu pulled down, indicating the two options available for selection: Inventory and Mesh. Note that the default setting is Inventory,

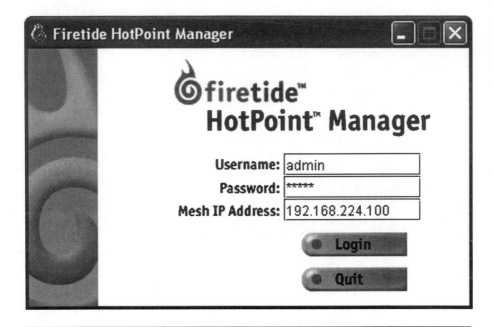

Figure 6.2   The log-in screen of the HotPoint Manager requires the user to enter a username, password, and Mesh IP address.

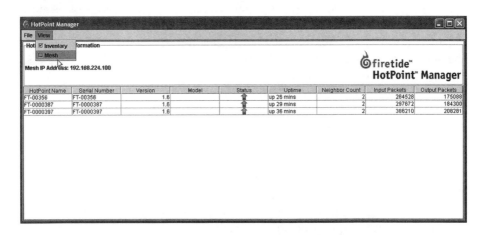

Figure 6.3   Obtaining an Inventory view of the mesh network through the use of the HotPoint Manager program.

which produces a tabular display as shown in Figure 6.3. The text obscured by the pull-down menu is "HotPoint Network Information" which represents the title for the Inventory screen display.

In examining Figure 6.3 note that the Inventory page has nine columns. The first column, HotPoint Name, is the unit's serial number which is used as a default. Thus, the entries in the first two columns are initially the same. If you want to rename a HotPoint router you could simply right-click on a name in the first column to enter a more intuitive name, such as "kfc roof", "conference room", or "office."

The third column displays the version of firmware used by each router. The fourth column, which is labeled "Model" was not used with the version of HotPoint Manager used by this author. The fifth column, which is labeled "Status," uses a green "Up" arrow to indicate that a HotPoint router is actively connected to the mesh. If you power off a previously viewed router, move the router out of the range of other routers in the mesh, or incorrectly position the router's antenna and lose connectivity to the mesh the symbol in the column will turn to a red "x." Thus, a simple glance at the Status column entries can provide you with an instant determination of the status of the nodes in your HotPoint mesh network. Continuing our tour of columns in the Inventory View menu, the Neighbor Count column indicates the number of other mesh nodes with which each HotPoint router has direct radio access. Because this author was using three HotPoint routers that could hear one another, each router could hear two neighbors, resulting in a neighbor count of two for each device.

The last two columns, labeled "Input Packets" and "Output Packets" provide a running total of packets received from and transmitted onto the mesh network. This count includes mesh routing traffic, HotPoint Manager data traffic, and user data traffic.

You can obtain a good indication of the overhead associated with mesh communications by powering up a series of HotPoint routers and examining their packet count field values prior to using the mesh network for moving data traffic. For example, from Figure 6.3 the first HotPoint router is shown to have an uptime of 26 minutes, which represents 26*60 or 1560 seconds. Because the router transmitted 175,088 packets during the uptime period, this works out to a packet transmission rate of 175,088 packets/1,560 seconds, or approximately 112 packets per second. Although the HotPoint Manager does not count octets or bytes, later in this chapter we examine the use of Wild Packets AiroPeek program to capture and decode packets transmitted between HotPoint routers. For now, if we assume 64-byte UDP packets convey intrarouter information, the overhead becomes:

$$\frac{112 \text{ packets/second} * 64 \text{ bytes/packet} * 8 \text{ bits/byte}}{11 \text{ Mbps}} = .005224$$

The 11 Mbps used in the equation's denominator reflect the highest IEEE 802.11b operating rate of the Firetide HotPoint routers. Thus, using 64-byte UDP packets for intrarouter communications requires approximately one-half of one percent of the wireless bandwidth. Thus, the overhead associated with intrarouter communications can be considered to be insignificant.

Mesh View

A second option in the View menu provides both a graphic view of your mesh network and the ability to control several configuration parameters via a series of menu tabs. That view is the Mesh View, which is illustrated in Figure 6.4. In examining the Mesh View shown in Figure 6.4, note it is subdivided into two areas: the upper window provides a graphic view of the nodes in the mesh network and the lower window provides you with the ability to view and change, if desired, a series of mesh network configuration parameters.

Setup Tab

When you select the Mesh View, the lower window displays the contents of the Setup tab. In Figure 6.4 the default settings associated with the Mesh IP address, username, and password are displayed. You can change any or all default values by typing a new entry into an applicable box. Once you click on the Save button the mesh network will reboot to activate the new settings, requiring a period of up to a few minutes during which all network operations will be suspended.

Security Tab

The second tab in the lower Mesh View window is labeled "Security." Its selection provides you with the ability to set and enable AES encryption and WEP security. Figure 6.5 illustrates the lower window of the Mesh View selection from the View menu, showing the Security tab parameters.

Firetide wireless mesh routers support 128-bit AES encryption of data traveling between nodes in the mesh. By default AES encryption is disabled. If you enable AES encryption you should change the default 32-hex character key as that key is printed in every HotPoint Manager user manual. Similarly, if you enable WEP you should change the 128-bit default key shown in the right portion of Figure 6.5. Similar to changing the previously covered Setup tab parameters, changing an encryption setting will result in the mesh rebooting. This activity can interrupt all mesh

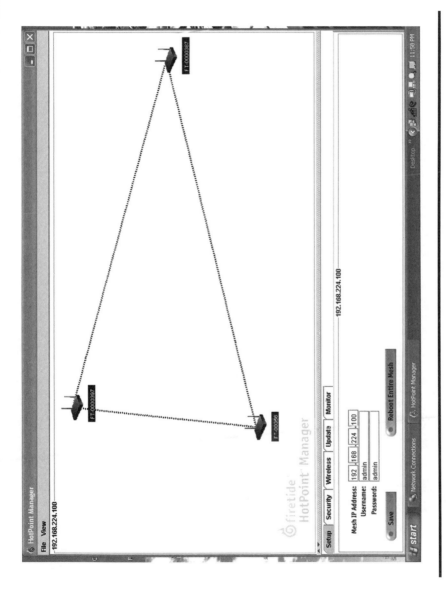

Figure 6.4 The Mesh View provides a graphic illustration of the hop relationship of routers in the mesh as well as the ability to view and change configuration parameters.

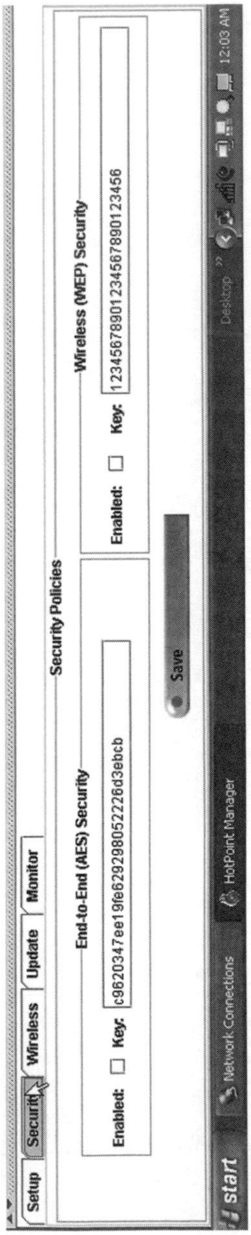

Figure 6.5 The HotPoint Manager security tab controls enabling AES encryption and WEP security to include specifying a key for each.

operations for up to a few minutes, during which the new security settings are distributed and loaded into other routers and those routers are restarted.

Wireless Tab

The Wireless tab controls the setting of the SSID or network name and the IEEE 802.11b channel used by the routers for communications.

Figure 6.6 illustrates the lower portion of the Mesh View screen when the Wireless tab is selected. Note that by default the SSID is set to "firetide" and channel 1 is used for mesh communications. Although there are 11 channels available for use by 802.11 wireless LANs, overlapping of channel frequency results in channel 1 being a good default value. This is because most access points by default use either channel 6 or channel 11, resulting in negligible interference between HotPoint mesh routers and any access points within the radio area covered by the mesh network. Concerning radio channels, Firetide reserved channel 11 for future use, resulting in limiting the selection of a radio channel to 1 to 10 for HotPoint routers.

Update Tab

The Update tab, which is illustrated in Figure 6.7, provides the ability to update HotPoint routers with the latest release of firmware updates. Through the use of the Update tab you would first select an update file previously downloaded from the Firetide Web site. Clicking on the Perform Upgrade button would update each of the HotPoint 1000S wireless mesh routers in the network in one operation, alleviating the necessity to individually upgrade each router.

Monitor Tab

The fifth tab, which is labeled "Monitor" when selected, provides a two-window view similar to that previously shown in Figure 6.4. However, the selection of the Monitor tab results in the tabular display of six columns of data for each of the HotPoint routers in the network being presented in the lower window. Table 6.1 lists the Monitor View columns displayed in the lower window as well as a brief description of the entry in each column.

In examining the entries in Table 6.1 it should be noted that the last four metric counters are reset to zero when a node reboots. This means you need to consider making a copy of the statistics prior to performing a firmware upgrade or another activity that could cause a reboot of the

Figure 6.6 The Wireless tab provides the ability to view and change the SSID and the operating channel the mesh network uses for communications.

Figure 6.7 The Update tab provides the ability to update firmware on all routers used in a Firetide HotPoint-based mesh network.

Table 6.1 Monitor Tab Display Columns

Column Label	Description
HotPoint Name	The name associated with each router, where the default is the router's serial number
MAC Address	The MAC address for the RF radio inside each router
No Route	The number of times a packet could not be transmitted due to the unavailability of a destination, such as when a node temporarily disconnects from the mesh
Fragment Timeout	The number of times a fragment of a large packet could not be received in a timely manner
Bad Checksum	The number of times a packet was received with a bad checksum

routers in the network. By having copies of the screen you can examine the performance of your mesh network over a prolonged period of time.

Creating the Indoor Mesh Network

As previously mentioned in this chapter, this author used three Firetide HotPoint routers to establish a mesh network in his home, By positioning two of the routers on the second floor he was able to overcome some of the prior wireless transmission limitations associated with a two-story stone fireplace in the middle of his home that attenuated RF signals between a wireless router located in his kitchen and certain rooms in the house.

The top portion of Figure 6.8 illustrates in schematic form the config uration of this author's Internet cable modem connecting his SMC Networks Barricade 802.11g wireless router and a Firetide HotPoint Wireless mesh router located on the built-in desk in the kitchen of his home. The lower portion of Figure 6.8 indicates the positioning in schematic form of two Firetide HotPoint routers in the second floor of the author's home.

In examining Figure 6.8a, as a review, note that the wireless operation of the SMC router occurred using channel 11 and the HotPoint router used its default setting of channel 1. This minimized any potential for RF interference between the two routers located next to each other.

The HotPoint router was cabled to an Ethernet switch port on the SMC router, with the latter cabled to the cable modem that terminated the cable connection to the Internet. Because the SMC router was configured to use its DHCP server capability, this action enabled up to 253 wireless and

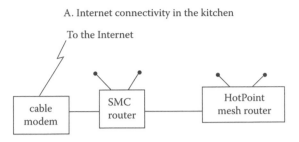

A. Internet connectivity in the kitchen

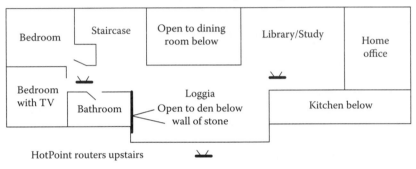

Figure 6.8   Positioning Firetide HotPoint routers in the author's home.

wired devices to become participants on the in-house network and share the single Internet connection.

The RF signals of the SMC router could be received in the author's upstairs bedroom. In his upstairs home office, however, signal strength was limited and adversely affected the data rate obtained from using a laptop and desktop computer. In addition, a trip with a laptop to either upstairs bedroom location to work and watch television was not productive because the two-story-high wall of stone in the author's den in effect blocked RF signals from the SMC router located in the first-floor kitchen area from reaching the upstairs bedroom areas. It should be mentioned that the bedroom with the TV shown in Figure 6.8b was more like a second den, because the author's children are grown and long ago moved into their own homes at the opposite end of the country. In fact, one bedroom has a desk, enabling this author to both use a laptop computer productively as well as view television if the laptop could only communicate with the downstairs SMC router. This communications requirement was satisfied by connecting a short straight-thru cable from the laptop to one of the three ports built into the HotPoint router located outside the

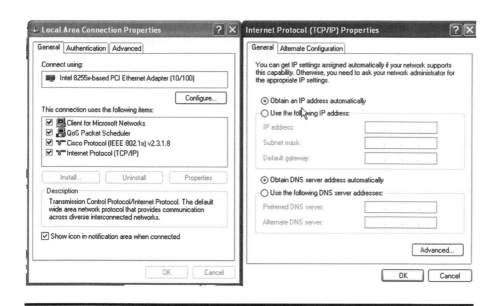

Figure 6.9 **The HotPoint mesh network provided a transport facility for obtaining an IP address from a DHCP server once the IP protocol Properties tab was set to "obtain an IP address automatically."**

door of the bedroom. That router's position is denoted by the router symbol positioned to the left of the loggia in Figure 6.8b.

Because the SMC router was configured to operate as a DHCP server the only configuration required to be made to the laptop was to use the Properties tab of the Ethernet adapter's TCP/IP protocol such that "obtain an IP address automatically" was selected as shown in Figure 6.9. Once this action was accomplished, the laptop used the three-router mesh network as a transport mechanism to the SMC router and through that route to the Internet. The use of the three HotPoint routers enabled this author to easily move his laptop between his home office and bedroom, requiring a simple cable connection to the closest HotPoint router to gain instant connectivity to the Internet. Because the second HotPoint router positioned on the second floor outside his home office was oriented more closely to the router on the first floor than the prior wireless connection between computers in the office and the SMC router, a higher data rate was obtained.

Considering a Community Network

Because it would be difficult to coordinate the placement of two Firetide HotPoint wireless mesh routers in other homes, this author established a community network within his home. However, the use of the HotPoint routers demonstrated the ease with which a true community network could be established. For example, using the outdoor version of the HotPoint router this author could strategically position the router on one of the many pine trees on his property. Because his neighborhood had underground utilities it would not be possible to coordinate the placement of HotPoint routers on utility poles. However, for neighborhoods that offer conventional utility poles it might become possible to contact the telephone or electric company to ascertain their position concerning the attachment of foreign equipment to their poles. Inasmuch as this was not a possibility, pine trees would be used in the author's neighborhood.

This would enable the author as well as other homeowners in the neighborhood to place HotPoint indoor wireless mesh network routers either inside their home at locations from which the indoor router could communicate with the outdoor router or cable their home computer to a nearby HotPoint outdoor router. For either situation a series of indoor HotPoint routers communicating with each other through the services of one or more outdoor routers would form a community network, with the Internet connection in the author's home being shared among a number of happy homeowners. Of course, because the cable network operator that provides Internet services for this author precludes this type of community Internet access sharing, a different type of Internet connection would be required. Other options that would allow the establishment of a community network with shared Internet access could include a commercial asymmetrical digital subscriber line (ADSL) or even a commercial satellite Internet access service.

The community network previously described can be extended to other environments. For example, consider a college campus or office park where buildings are dispersed over a considerable geographic area. Through the strategic placement of HotPoint indoor and outdoor routers within buildings as well as on light poles throughout the campus or office park it becomes possible to easily establish a community network to serve a large group of users. Because an average Web page consists of approximately 50,000 characters or 400,000 bits, the IEEE 802.11b wireless LAN transmission used by HotPoint routers can easily enable the display of over 600 Web pages per minute by all users of the mesh if we assume that only half of the 11-Mbps transmission capacity of the mesh network is available for use. Thus, well over 100 online users could be supported by a single mesh network if we assume that such users take just ten seconds to read or scan the contents of a Web page prior to clicking to a new page. Because most Internet users perform other activities, it's probably safe to assume that a much larger base of users could be supported by a single mesh network.

Because IEEE 802.11b wireless devices can operate in three nonoverlapped channels it's possible to create three nonoverlapping RF mesh-based networks within a common geographic area. This should provide sufficient support for satisfying the network requirements of many colleges and universities as well as office park employees and homeowners in various suburban and even urban neighborhoods. Now that we have an appreciation for the manner by which Firetide HotPoint routers can be used in a variety of network operations, let's turn our attention to the manner by which those routers communicate with one another. In doing so this author used another laptop computer operating WildPackets AiroPeek wireless LAN packet capture and decoding program to observe RF transmission both from his SMC Barricade wireless router and the Firetide HotPoint wireless mesh routers operating in his home. Because we are focusing our attention on the wireless mesh routers in the next section we limit our attention to observing and discussing the contents of packets flowing between the HotPoint routers.

6.3 Observing Intramesh Communications

Previously in this chapter we noted that this author was able to set up three Firetide HotPoint wireless mesh routers with a minimum of effort to establish a mesh network. In fact, if security were not an issue, one could use the HotPoint routers straight out of their boxes as they have the ability to automatically form a mesh network without requiring any configuration changes to existing router parameters. Because it's both interesting and informative to observe the inner workings of wireless devices by examining their RF signals, this author decided it would be worthwhile to use a packet decoder to examine their over-the-air transmission. Thus, this author used the WildPackets AiroPeek packet capture and decoding program to examine the over-the-air transmission occurring when the SMC Networks and Firetide HotPoint routers were operational.

AiroPeek Operation

Figure 6.10 illustrates the initial AiroPeek screen display after the Capture icon was selected, resulting in the Capture Buffer Options window displayed in the foreground of the screen. In the background you can observe that the AiroPeek display is subdivided into three windows. The larger window, which is currently blank, will display packets captured. In the lower portion of the screen there are two windows. The leftmost window

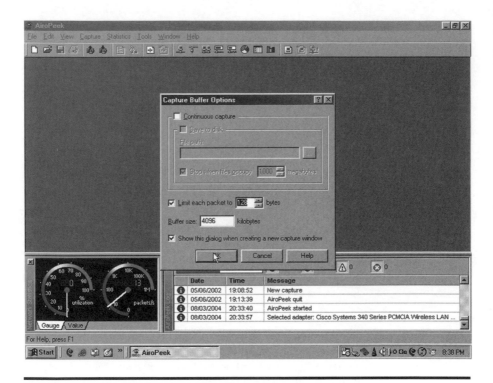

Figure 6.10   The initial WildPackets AiroPeek screen display showing available capture buffer options.

shows two gauges, indicating the utilization and packet rate of the monitored wireless LAN. By default, gauges are displayed, however, clicking on the label "Value" will result in the display of numerics. The third window, which is shown in the lower right portion of the screen indicates the use of the AiroPeek program, displaying such information as when the program started and terminated as well as the selection of a particular wireless adapter for use by the program and its data capture operations. From that window note that a Cisco Systems 340 series wireless LAN adapter was being used on the hardware platform running AiroPeek.

Turning our attention to the foreground window labeled "Capture Buffer Options," note the setup of the program for packet capture. Instead of using the default limit of 64 bytes per packet, the default was reset to 128 bytes, however, the buffer size default value of 4096 kbytes was retained. This setting permits a reasonable amount of packets to be captured that are of sufficient length to denote the wireless activity performed by the HotPoint routers.

Once the capture buffer is configured, packet capturing can occur via the capture menu, resulting in the display of a button labeled "Stop

Capture" in the upper right corner of the screen. That button's label will toggle between "Stop Capture" and "Start Capture" with the label's wording based upon the prior state of the buffer capture operation. In addition, the upper window will be split into two areas. The upper area of the top window will provide a summary of packets received, filtered, and processed as well as buffer memory usage. Because the proof of the pudding is in the eating, let's examine the AiroPeek screen display after a short packet capture session occurred.

Examining Captured Packets

Figure 6.11 illustrates the AiroPeek display after 155 packets were captured and stored in the memory of the computer on which AiroPeek operated. As previously discussed, the upper portion of the AiroPeek display is subdivided, with the main portion of the screen showing information about captured packets and the lower portion providing information about the cumulative number of packets received, filtered, and processed as well as computer memory usage. In examining the main portion of the upper window of the AiroPeek display, note information about the source

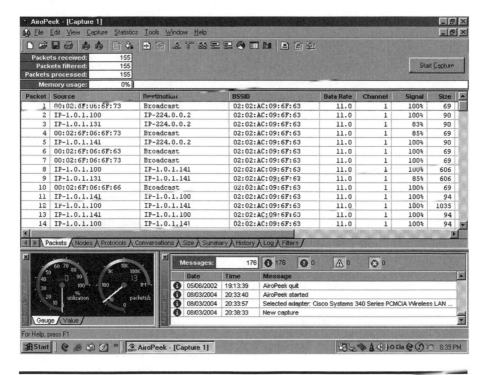

Figure 6.11 The AiroPeek display after a short packet capture operation.

and destination addresses in each captured packet, the Basic Service Set IDentifier (BSSID), data rate, frequency channel, signal strength associated with the captured packet, and size of the packet in bytes. Simply scrolling to the desired packet and double-clicking on a row results in the decoding of the packet.

A quick glance at the source and destination columns in Figure 6.11 indicates some interesting aspects about the HotPoint routers. If you first look at the source column you will note three unique MAC addresses, each corresponding to a router. In the destination column you will observe that the MAC source addresses are transmitting broadcast frames. Because this author was using three HotPoint routers, the prior description should come as no surprise inasmuch as each router periodically broadcasts information about itself as layer 2 frames. Now let's turn our attention to layer 3 addresses shown in the source and destination columns in Figure 6.11.

If you carefully examine the source and destination columns shown in Figure 6.11 you will note four IP addresses, however, as previously mentioned only three HotPoint routers were used to form the mesh. Although outwardly the routers form a mesh network that appears as a layer 2 switch, within the mesh the routers appear as an interconnected network. Thus, the mesh itself has an IP address and each router also has a unique IP address, resulting in four IP addresses being used by the three routers. Returning our attention to Figure 6.11, packets transmitted with any one of the three IP addresses of 1.0.1.x that communicate with the IP address 224.0.0.2 are in effect communicating with the mesh.

Packet Decoding

To illustrate the packet decoding capability of AiroPeek as well as to obtain additional insight into the radio transmission occurring let's double-click on the first packet captured that was positioned at the top of the upper window shown in Figure 6.11. The upper portion of the top window displayed by AiroPeek now shows the decoding of the over-the-air transmitted packet. In actuality, the captured transmission represents an IEEE 802.11 wireless frame because all packets are encapsulated by wireless Ethernet frames for transmission over a wireless network. However, because most publications refer to transmission in terms of packets we continue in this tradition even though it is technically incorrect. The only exception to this occurs when we refer to IEEE 802.11 functions, for which the term "frame" is used.

In examining a partial decode of the captured packet shown in Figure 6.12 note that the 802.11 MAC header indicates the frame type as "Management" and its subtype is "Beacon." Thus, prior to examining the frame a few words are in order concerning operation and utilization.

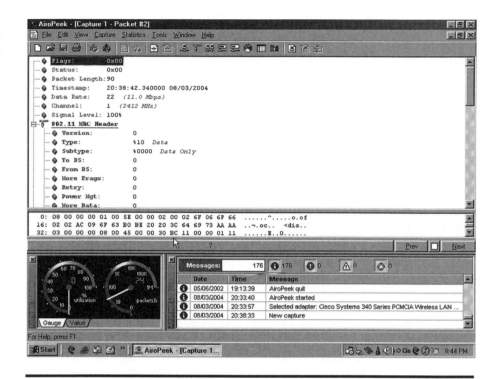

Figure 6.12 The initial decode of the second packet indicates it is a data packet.

The Beacon Frame

A Beacon frame is periodically transmitted to announce the presence of an access point and relay such information as the BSSID of the network to other stations.

If you return your attention to Figure 6.11 and examine the packets with a destination address of "Broadcast" you will note three MAC addresses in the "Source" column associated with broadcast packets. Those MAC addresses correspond to the three HotPoint wireless mesh routers that were used to form the mesh network previously described in this chapter. Returning our attention to Figure 6.12, in the upper portion of the referenced illustration, only a few bytes of the actual decoding are shown. To see the rest of the decoding you would need to either scroll down the upper window, use the File menu to save the decode to a file, or focus your attention upon the hexadecimal decode shown in the lower window. Prior to focusing our attention on the last-named, a quick review of the format of a wireless frame is in order.

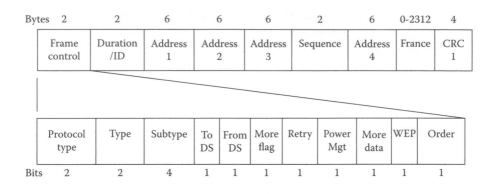

Bytes 2 / 2 / 6 / 6 / 6 / 2 / 6 / 0-2312 / 4

Frame control	Duration /ID	Address 1	Address 2	Address 3	Sequence	Address 4	France	CRC 1

Protocol type	Type	Subtype	To DS	From DS	More flag	Retry	Power Mgt	More data	WEP	Order

Bits 2 / 2 / 4 / 1 / 1 / 1 / 1 / 1 / 1 / 1 / 1

Figure 6.13   The general format of IEEE 802.11 frames.

Wireless Frame Format

Figure 6.13 illustrates the general format of an IEEE 802.11 wireless media access control frame. It should be noted that there are several types of MAC frames, resulting in portions of certain fields either not used or not included in certain types of wireless frames.

If we look at both Figure 6.12 and Figure 6.13 we can make better sense out of the hex decode shown in the middle window of Figure 6.12. For example, the fifth byte of the decode is shown as "FF." That setting of hex "FF" continues for five more bytes, resulting in bytes 5 through 10 of the frame being set to all 1s. This corresponds to placing a broadcast (all binary 1s) in the Address 1 field. Because Address 1 is always the recipient address, in effect, the Beacon frame is being broadcast to all devices. The following 6 bytes, hex values 00 02 6F 06 6F 73 represent the value of the Address 2 field, which is the station that transmitted the Beacon. Because the frame was not transmitted over a distribution system (DS) inasmuch as the To DS and From DS bits are not set, Address 3 represents the BSSID. Thus, the next six bytes are shown as having the hex value 02 02 AC 09 6F 63 which corresponds to the BSSID value shown in the column labeled "BSSID" for packet 1 in Figure 6.11.

If we scrolled down the upper window of Figure 6.12 we would observe more entries in the 802.11 MAC Header decode, including Address 3 which represents the BSSID followed by two bytes of Sequence control information, followed by Address 4, and so on. The sequence control field is used to represent the order of different fragments belonging to the same frame as well as to provide a mechanism to recognize packet duplication. This two-byte field consists of two subfields, fragment number and sequence number, which define the frame and the number of the fragment in the frame. From the hex decode window shown in Figure 6.12 the Sequence Control Field is shown to have a value of B0 36.

The following field is the Address 4 field. Address 4 is used for the special case where a wireless distribution system is employed, resulting in a frame being transmitted from one access point to another. In such cases the To DS and From DS bits in the Frame Control Field should be set according to IEEE 802.11 standards. In the 802.11 MAC header shown in Figure 6.12 note that neither of those two bits are set, however, the hex decode shown in the middle window of that illustration indicates a value for address 4 of hex 98 11 4E 6A 00 00. Although a beacon transmitted by an access point would, under the IEEE 802.11 standard, set this field to all 0s when the To DS and From DS bits are both set to 0 values, Firetide HotPoint routers do not do so. Instead HotPoint routers appear to use the Address 4 field in a proprietary manner.

As part of the Beacon frame, the two fields following the Address 4 field contain the Beacon interval and a timestamp. The Beacon interval represents the amount of time between beacon transmissions. The timestamp field value is used by each receiving station to update its local clock, enabling synchronization among all routers in the mesh.

Continuing our examination of the decoding of packet 1, the data field or frame body follows the Address 4 field. If you again focus your attention upon the hex decode shown in Figure 6.12 you will note that the ninth through sixteenth bytes following the Address 4 decode are hex values 66 69 72 65 74 69 64 65. As indicated by the English translation on the third line to the right of the hex decode, those hex values correspond to the term "firetide" which represents the HotPoint wireless mesh router default SSID value. Thus, the SSID value of "firetide" represents the network name or service set identifier, and the BSSID of 02 02 AC 09 6F 63 represents the cell identifier whose value is the same as the MAC address of the radio and is transported in the Address 3 field in the frame.

There are several additional fields in the Beacon frame that deserve mention. The SSID field is followed by a Supported Rate field and Parameter Set field. The Supported Rate field indicates the data rates supported by the device transmitting the Beacon frame. For example, because HotPoint routers support the IEEE 802.11b standard, they can operate at 1, 2, 5.5, or 11 Mbps. Thus, a Beacon frame would indicate which data rates the device supports. The Parameter Set field indicates information about the wireless LAN support of the device, such as the channel number it is using.

Because Firetide HotPoint routers operate in ad hoc mode there is no access point in a HotPoint wireless mesh network. Instead of an access point transmitting beacons, one of the HotPoint routers assumes the responsibility for beacon transmission. After receiving a beacon, each station waits a period of time equal to the beacon interval and then transmits its own beacon if no other station does so after a random time delay. This action ensures that at least one station transmits a beacon and the random delay rotates the responsibility for transmitting beacons. Thus, if you return your attention to Figure 6.11 and examine the Source column for each Destination column "Broadcast" entry, you will note three distinct MAC addresses. Those addresses correspond to the MAC addresses of the three HotPoint routers used to form the wireless mesh network.

Now that we have an appreciation for the general decoding of a Beacon frame let's continue our decoding effort. In doing so let's move to the second packet shown in Figure 6.11 which has a source IP address of 1.0.1.100 and a destination IP address of 224.0.0.2.

Data Frame Decode

HotPoint routers use TBRPF, a proactive routing algorithm to route traffic. Within the mesh formed by the routers they automatically use hidden IP addresses that are different from the default IP address used with the HotPoint Manager program. Those hidden IP addresses are observable from the use of AiroPeek or a similar packet capture and decoding program, with the use of those IP addresses transparent to the flow of data through the mesh network. This enables DHCP, DNS, and other services provided by an access point connected to one HotPoint router to flow through the mesh and be received by clients connected to other HotPoint routers.

To illustrate the internal use of IP addresses by HotPoint routers let's turn our attention to the second packet that was captured in Figure 6.11. That packet is shown as having a source IP address of 1.0.1.100 and a destination IP address of 224.0.0.2. Both of those IP addresses are preassigned to the HotPoint routers and cannot be reconfigured.

The initial portion of the decode of the second packet shown in Figure 6.11 is displayed in Figure 6.13. If you focus your attention upon the portion of the decode labeled "802.11 MAC Header" and examine the Type and Subtype field values you will note that the decode indicates that the second packet represents a data packet. To obtain additional information about this data packet we need to scroll down the top window of Figure 6.13, so let's do so.

IP Header

Scrolling down to the beginning of the IP header results in the display shown in Figure 6.14. In examining the decode of the IP header we can note that IP is transporting UDP, because the protocol field value is set

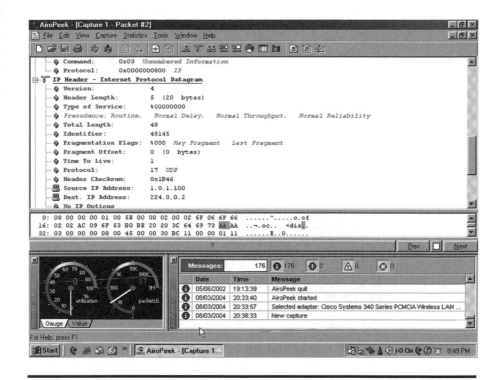

Figure 6.14   Examining the IP header of the second packet indicates it flows from a public IP address to another public IP address and transports UDP.

to 17. We can also note that the source IP address is 1.0.1.100 and the destination IP address is 224.0.0.2. Because the address of the mesh network is 224.0.0.2, in effect, a router whose IP address is 1.0.1.100 is transmitting data to the mesh network.

If we compare the source and destination IP addresses to RFC 1918 address blocks we would note that those addresses are not contained in those address blocks. Thus, the HotPoint routers use public instead of private IP addresses. We can obtain additional information concerning the data packet by continuing to scroll down the top window which will then display the decode of the UDP header because Figure 6.14 indicates that the IP datagram is transporting a UDP segment. Thus, let's continue our examination of packet header information by continuing our scroll down through the top window in the AiroPeek display.

UDP Header

Once again we scroll down the top AiroPeek window that shows the decode of packet number 2. This time we do so to ascertain the application

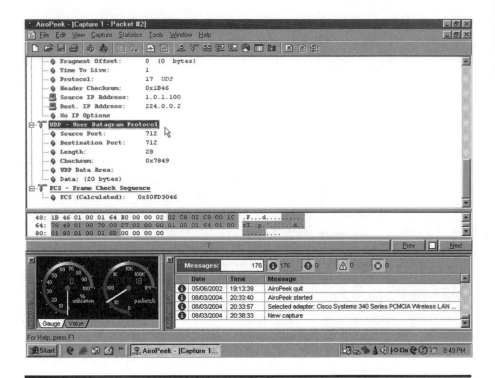

Figure 6.15 **The UDP header indicates that the data packets transports TBRPF routing information.**

transported by UDP. As a refresher, the Protocol field in the IP header indicates the protocol transported, which when set to a value of 17 indicates UDP. When attempting to determine the application transported by UDP we need to examine the destination or port field value in the UDP header,

In examining the UDP header decode shown in Figure 6.15 note that both source and destination field values are set to 712. That port number is used to identify the TBRPF routing protocol. Thus, packet number 2 represents an IP datagram consisting of a UDP segment that transports TBRPF routing protocol information. This means that when the HotPoint routers are powered on, they broadcast beacons as well as automatically exchange routing information using TBRPF, all without requiring user intervention. This explains why, to the end user, HotPoint routers are in essence plug-and-play devices that can be used straight out of their boxes.

Chapter 7

Wireless Mesh Standards

One of the major limitations of wireless mesh networks is a lack of standardization. Currently, vendors offering wireless mesh networking products employ different routing algorithms as well as support different transmission platforms, such as the IEEE 802.11a, b, and g protocols. In addition, several vendors modified the operation of the 802.11 MAC (Media Access Control) layer. This lack of standards means most mesh network implementations are single-vendor solutions and this obviously ties the network operator to a single vendor. If that vendor raises equipment prices, goes out of business, or merges with a different vendor the operator of the mesh network can experience a significant change in equipment availability and network support. In addition, the lack of current standards more than likely adversely affects the efficiency of IEEE 802.11-based mesh networks. The reason for the latter results from the fact that the 802.11 protocol was not designed with a mesh topology as a primary consideration.

As mesh networking standards evolve it's possible that a method will be defined that enables the capture and distribution of network statistics throughout a mesh network. This information could then be used by the upper layer routing protocol to enhance the movement of data across the mesh network.

In this chapter we first turn our attention to the evolution of mesh networking standards, discussing some of the more common routing protocols as well as vendor efforts toward standardization of the technology. In the second section in this chapter we briefly discuss the emerging IEEE 802.11s standard. Although possibly two to three years away from being finalized, this standard, when put into effect, offers the promise to considerably expand the use of mesh networks in the same manner that the IEEE 802.11a, b, and g standards expanded the growth of wireless LANs.

7.1 Evolution

The evolution of mesh networking standards can be considered to have begun with the selection of a transmission platform and routing protocol.

Transmission Platform

During 2003 and moving forward into 2004, several pioneering wireless mesh networking equipment manufacturers selected the IEEE 802.11 series of transmission protocols as the basis for the development of mesh networks. Unfortunately, some vendors slightly modified the operation of the MAC layer of the transmission protocol they selected for use. This action negated the benefits of using a transmission standard and represents a key reason for the need for a standard.

Routing Protocols

In the area of mesh routing protocols the area of potential standardization is currently very fragmented. Over the past decade papers were published describing the operation of approximately 50 routing protocols. Although many of those protocols were implemented in equipment for laboratory experiments, only a few were actually used in commercial equipment at the time this book was prepared. To further compound the standardization problem, vendors modified the operation of some routing protocols in an attempt to better tailor the operation of the routing protocol they selected for use in their equipment.

Out of the approximately 50 routing protocols developed, there are 4 that are candidates for potential standardization in an evolving wireless mesh network standard. A brief description of each mesh routing protocol is contained in Table 7.1. Of the four listed in the table, AODV and TBRPF in this author's opinion are most likely to win the mesh network routing protocol war and were described earlier in this book. Because AODV and TBRPF have strengths and weaknesses with respect to the size of the mesh in terms of nodes supported, this author believes that the evolving IEEE 802.11s standard that will define the operation of mesh networks may very well allow the use of either routing protocol, enabling end users

Table 7.1   Mesh Routing Protocols

Protocol	Protocol	Description
AODV	Ad hoc On-demand Distance Vector	Developed for mobile ad hoc networks. Has low overhead and is in the public domain.
DSR	Dynamic Source Routing	Supports self-organization and self configuration. Each data packet contains full route information. Works well in smaller stable networks.
ODMRP	On-Demand Multicast Routing Protocol	A multicast on-demand routing protocol developed for ad hoc networks. Using on-demand procedures, it dynamically builds routes and maintains multicast group memberships. Protocol is well suited for limited bandwidth, frequently changing topology.
TBRPF	Topology Broadcast Based on Reverse-Path Forwarding	Proactive routing protocol that uses link-state routing and provides hop-by-hop routing along minimum hop path.

to select a protocol that operates more efficiently with respect to the number of nodes in the network

Standards Responsibility

When discussing mesh networking standards it's important to understand the differences between the first two layers and the third layer in the protocol stack used by mesh networking devices. The first two layers are the PHY (Physical) and MAC layers. For products that use an IEEE 802.11 transmission platform, the IEEE defines the standards for those layers. The actual routing protocol, such as AODV or TBRPF represents a layer 3 Internet Protocol (IP) operation in the protocol stack and is not the responsibility of the IEEE. Instead, layer 3 IP protocols are the responsibility of the Internet Engineering Task Force (IETF). The IETF standardization process results in the issuance of Request for Comments (RFCs) which move from a draft to a published standard.

Recognizing the need for standards covering wireless mesh networking, Cisco Systems and Intel proposed a new standard for wireless mesh networking at an IEEE meeting in Vancouver, Canada, in January, 2004. The effort of Cisco and Intel as well as the recognition that the lack of interoperability was limiting the acceptance of mesh networking products were more than likely contributing factors in the drafting of a Project Authorization Request (PAR) for forming an IEEE 802.11 Extended Service Set (ESS) Mesh Task Group within the IEEE 802.11 Working Group. Approval of the PAR will eventually result in a new addition to the IEEE 802.11 series of standards, with the 802.11s standard nomenclature reserved for a wireless mesh local area network standard. Thus, in the remainder of this chapter we turn our attention to the previously mentioned PAR and the goal of the 802.11s standard that will emerge from the work of the Mesh Task Group. However, as previously discussed, because the IEEE's responsibility for standardization is limited to the lower two layers in the protocol stack, the IETF will be responsible for the standardization of routing protocols that will, in effect, be transported via the evolving IEEE wireless mesh standard.

7.2 The Proposed IEEE 802.11s Standard

The PAR for forming an ESS Mesh Task Group within the IEEE 802.11 Working Group as with similar 802.11 standards is concerned with the two lower layers in the Open System Interconnection (OSI) Reference Model. Thus, the resulting 802.11s standard will eventually define PHY and MAC layers for mesh networks that will enable an improvement in their area of coverage with no single point of failure.

From an examination of the PAR, the expected date of submission for an initial ballot is shown as January 1, 2006, with a project completion date indicated as January 1, 2007. Thus, if the Mesh Task Group meets its schedule the promulgation of a wireless mesh networking standard is at best several years away.

Objective

The objective of the proposed project is to develop an IEEE 802.11 ESS mesh using IEEE 802.11 MAC and PHY layers that support broadcast, multicast, and unicast delivery of information over self-configuring multi-hop topologies. The effort of the Mesh Task Group will result in an extension to the current IEEE 802.11 MAC layer, defining an architecture and protocol that enable an 802.11 ESS mesh to be constructed and enable its efficient operation.

The 802.11 ESS mesh will be created via a collection of access points interconnected via wireless transmission that enables automatic topology learning and dynamic path configuration to occur. In addition, the resulting mesh network will use the 802.11 MAC to create an 802.11 wireless distribution system that supports unicast, multicast, and broadcast delivery at the MAC layer through the mesh.

Under the explanatory notes portion of the PAR it's mentioned that the ESS mesh shall be extensible to "enable alternate path selection metrics and/or protocols based on application requirements." Thus, it appears statistical information about path utilization will be distributed and the lower two layers will support a variety of routing protocols. Also included in the explanatory notes is a reference to a target configuration of up to 32 devices participating as AP forwarders in the ESS mesh, and larger configurations may also be contemplated by the standard. Thus, the ultimate maximum size of a mesh network is unclear at this time. In actuality, because an ESS mesh will interface with higher layers in the protocol stack, a combination of nodes and the higher-layer protocols in use will more than likely govern the ultimate size of the mesh.

Security

A wireless mesh network is similar in scope to a wireless LAN in that radio transmissions can be intercepted. Thus, it should come as no surprise that the previously mentioned PAR includes the mention of security. In the PAR it's stated that "IEEE 802.11i security mechanisms, or an extension thereof" will be used to secure an ESS mesh. What this indicates is that at the present time the exact method that will be used for securing an ESS mesh may evolve over time as the 802.11s standard evolves.

Project Status

Since the publication of the previously mentioned PAR, the ESS Mesh Task Group has had two meetings. The first was held during July, 2004 in Portland, Oregon, and the third occurred during September, 2004 in Monterey, California. At the July meeting the Task Group decided upon priorities for the September meeting as well as discussed such important issues as security, quality of service, routing, and the agreement of basic definitions. The second meeting during 2004 included a number of presentations of various aspects associated with mesh networking and illustrated why the development of a standard may very well require the full duration of the PAR period. Issues including support of QoS to enable VoIP to be transported over the mesh as well as alternate path usage and the support of a large number of nodes will require a considerable amount of effort.

Chapter 8

The Future of Wireless Mesh Networking

In this concluding chapter this author polishes his crystal ball to predict the future of wireless mesh networking.

In previous chapters we examined the IEEE series of wireless LAN standards, more formally referred to as the IEEE 802.11a, b, and g standards to obtain a foundation for examining wireless mesh networking. In succeeding chapters we examined the common protocols used for routing at the network layer, examined in detail how one vendor's product operated within this author's home as a basis for extending a mesh to cover a wide geographical area, and discussed the evolving IEEE wireless mesh networking standard. Unfortunately, we noted that that standard is several years away from promulgation.

Prior to making a prediction a few words are in order concerning several important areas. First, we need to obtain an appreciation for the reason a lack of standards holds back the broad adoption of the technology. Next, we need to obtain an appreciation for the economics associated with providing equipment to the current main categories of adapters of the technology. Once this is completed we look at the positive market drivers for the technology that will enable this author to make a rational prediction.

8.1 Standardization Problems

Because current wireless mesh networking products are proprietary there is a reluctance among business organizations and federal government agencies to establish mesh networks. This probably explains why the majority of adapters of this technology fall into two or three categories.

The first category is small municipalities that for various reasons either cannot obtain wideband Internet access at a reasonable price or were bypassed by one or more communications carriers when the broadband infrastructure in the form of orange fiber-optic cable was rapidly installed to link major metropolitan areas around the globe during the late 1990s through the turn of the new millennium.

The second category of wireless mesh networking adapters is small companies that are very cost conscious and, due to their size, the fact that they are locked into a proprietary solution is of little consequence. In comparison, a large organization that is considering a trial of wireless mesh networking products will more than likely be hesitant to proceed because a positive trial that would lead to a corporate rollout would force the company into a single-vendor solution. As many readers are well aware, single-vendor solutions based upon proprietary products can result in significant problems if the vendor exits the market, decides to scale back support, or initiates an across-the-board price increase for the product or its support.

The third category of wireless mesh networking adapters is the hobbyists who are constructing mesh networks, commonly in rural areas to enable their neighbors to share a common broadband Internet connection.

Economics

If we think about the market provided by the three categories of wireless mesh networking products mentioned previously in this chapter we can note they represent a minor or niche product market. In fact, during this author's research effort performed during the development of this book it was noted that the major product vendors developing mesh networking products for the most part are startup companies with a minimal sales history. That said, let's examine some of the potential positive drivers that could shape the market for wireless mesh networking products.

Market Drivers

There are three key market drivers for wireless mesh networking products. Those drivers include the standardization of the IEEE 802.11a, b, and g technologies, the pending, albeit two years forward, standardization of wireless mesh networking products and the growth in the number of municipalities considering providing hot spots of Internet access. Let's discuss each driver to obtain an appreciation as to how they could affect the market for the technology.

IEEE 802.11a, b, and g Technologies

The IEEE 802.11a, b, and g wireless LAN platforms represent the most rapid deployment of any LAN technology, wireless or wired. Since the introduction of the IEEE 802.11 standard, the a, b, and g extensions resulted in over 10 million wireless LAN adapters and a million access points being shipped each year for the past two years. Because most vendor products although proprietary in design use the IEEE 802.11a, b, or g standard as a transmission base, this means there is an enormous base of equipment that could be integrated into wireless mesh networks.

IEEE Standardization of Mesh Networking

As mentioned both earlier in this book and briefly in this chapter, the IEEE is working on the development of a standard for wireless mesh networking. Assuming the standard can be implemented via a firmware upgrade into existing IEEE 802.11 series products, this means that once the standard is promulgated tens of millions of computers with existing IEEE-compliant wireless LAN adapters would also become mesh-network-enabled products.

Municipal Hot Spots

One of the more interesting aspects in the growth of Internet hot spots has been municipalities that are starting to offer expanded hot zones that cover the downtown of a small city. Although some of the first adapters of hot zones were small municipalities, just recently the city of brotherly love, better known as Philadelphia, was considering providing Internet access to all residents of the city.

A variety of solutions can be used to provide Internet access for residents of municipalities, however, the use of a mesh network probably provides a more cost-effective solution than the use of WiMax or the establishment of access points based upon the IEEE 802.11 series of standards.

Now that we have an appreciation for both existing mesh networking problems and potential market drivers, this author can dust off his crystal ball and make an educated prediction that readers can understand the basis for the prediction.

8.2 Mesh Networking Prediction

For the next two and a half to three years this author predicts that wireless mesh networking products will continue to represent a niche market, evolving gradually but without a significant increase in utilization. Approximately two years from now the IEEE's effort toward the standardization of wireless mesh networking should be finalized. Assuming mesh networking product vendors incorporate the standard into their products within a period of six months and individuals can obtain firmware upgrades to their existing IEEE 802.11a, b, and g wireless LAN adapters, all the forces will be in play to enable the large-scale adoption of the technology. Thus, it is this author's prediction that wireless mesh networks will reach a level of wide-scale deployment within the next two and a half to three years. This deployment will result in many municipalities providing their taxpayers with broadband Internet access that could significantly affect the revenue stream of Baby Bells that provide DSL service and cable television providers that offer broadband Internet access via cable modem connections. Thus, for the next few years, the use of wireless mesh networking products will be limited to experiencing a moderate level of growth. However, upon the standardization of the technology, the primary impediments that constrict the potential of the technology will be eliminated and the growth in the level of products shipped and upgraded via firmware can be expected to represent a significant market.

Index